"A timely, practical, and insightful book on the psychology of climate action for anyone who wants to feel more hopeful and be more courageous in the face of what is probably the biggest challenge of our times."

Professor Martijn van Zomeren, *University of Groningen (2024)*

"I strongly recommend this excellent and thought-provoking book. It meets many needs: partly a crash course in the theory of psychology and motivation, partly a self-help guide for sustaining strength and courage, and partly a strategic manual for social change activists. Above all, it recognizes that change originates in collective identity and that activists will win when they nurture and sustain that sense of joyful belonging."

George Marshall, *author of* Don't Even Think About It: Why Our Brains Are Wired to Ignore Climate Change *(2024)*

The Psychology of Collective Climate Action

How do we find the courage to act together against the climate crisis? This book weaves together real-life findings and examples from the socio-ecological movement with psychological research to show how motivation for collective climate action can be built.

The book addresses two key questions: how can individuals be motivated to participate in collective climate action, and how can climate groups become resilient and effective? Specifically, it explores how individuals can foster their identification with climate action groups and the belief in their joint efficacy. It touches on a wide range of topics, covering anger, moral considerations, activist burnout, and the perception of protests, as well as general theories of socio-ecological change.

This book is for anyone who is seeking the courage to act together and is curious about psychological insights. It will be essential reading for climate and environmental practitioners, climate activists and campaigners, climate change communicators, and anyone involved in socio-ecological change. It will also be of interest to students and researchers in the fields of environmental psychology, climate change, collective action, and political psychology.

Karen Hamann is a postdoctoral research fellow at the Department of Social Psychology of Leipzig University, Germany. Her research concerns empowerment, collective action, visions of an ecological future, and the energy transition. She is a co-founder of Wandelwerk e.V., a collective of practice-oriented environmental psychologists of which all the authors are members.

Eva Junge works as an environmental psychologist and climate communicator. She is a co-founder of Wandelwerk e.V., an advisor for socio-ecological NGOs, and an "artivist".

Paula Blumenschein is a research associate at TU Dortmund University, Germany. She works in the clinical and biological psychology working group with a focus on the climate crisis and mental health.

Sophia Dasch is an environmental psychologist and science communicator, specializing in climate education and consultation. She also works as a project manager at ConPolicy, an institute for consumer policy.

Alex Wernke works as a climate communication trainer for klima*kollektiv and as a freelance political educator.

Julian Bleh is a research associate at the Department of Social Psychology, Leipzig University, Germany, where his research is concentrated on the ability to envision social change.

The Psychology of Collective Climate Action

BUILDING CLIMATE COURAGE

Karen Hamann, Eva Junge,
Paula Blumenschein, Sophia Dasch,
Alex Wernke and Julian Bleh

Routledge
Taylor & Francis Group

LONDON AND NEW YORK

Designed cover image: Cover illustration by Otto Barboni (CC-BY 4.0)

First published 2025
by Routledge
4 Park Square, Milton Park, Abingdon, Oxon OX14 4RN

and by Routledge
605 Third Avenue, New York, NY 10158

Routledge is an imprint of the Taylor & Francis Group, an informa business

British Library Cataloguing-in-Publication Data
A catalogue record for this book is available from the British Library

ISBN: 9781032905297 (hbk)
ISBN: 9781032905280 (pbk)
ISBN: 9781003558439 (ebk)

DOI: 10.4324/9781003558439

Typeset in Palatino LT Std
by Newgen Publishing UK

For those committed to social change and climate justice, especially to our friends and colleagues at the change collective Wandelwerk. Thank you.

For Maël, may the climate movement, supported by the insights of this book, provide a sustainable future for you and all the children across the globe.

CONTENTS

FOREWORD

I am delighted to write a foreword for this important and insightful book on the psychology of climate protest and engagement. As the authors note, acting against climate change in essence requires courage, and I believe this book offers practical pathways towards being courageous together. These pathways are based on decades of academic theorizing and research, which means they are in line with a modern scientific understanding of the psychology of climate action.

The authors have also done a great job at translating these insights into concrete, actionable recommendations and guidelines. As such, the book uniquely brings together the psychological science of what moves and motivates people for climate action with the urgent need to act together. I can recommend the book to anyone who wants to feel more hopeful and be more courageous in the face of what is probably the biggest challenge of our times.

– Professor Martijn van Zomeren, University of Groningen (2024)

We live in an exceptional time in human history, and climate activists struggle with three huge challenges – their knowledge of the reality of what lies ahead, the fear that, like Cassandra, they shout that truth but are not heard, and the sense of responsibility that the fate of future generations depends on them.

This book is important because it understands those psychological challenges and addresses them directly. Wisely, it recognizes that engaging people in climate action is ultimately not about educating people – of producing yet more reports about parts per million of gases or degrees of temperature – but about investing in shared identity and progressive values, working through trusted networks and giving people a sense of belonging and collective agency.

It shows how a strong sense of community and shared values can sustain us, and, in turn, suggests ways that this community can be nurtured. The book is very strong on theory and source material, yet also provides very good practical advice. Of especial value to activists, it recognizes that they are vulnerable – that burnout is a real danger – and provides good advice on sustaining their activism – or, as the book says, building courage. I wish we had had this kind of material when I started my climate activism 30 years ago!

– George Marshall, environmental campaigner, communications
specialist, and founder of Climate Outreach (2024)

PREFACE

CLIMATE COURAGE AND THE AIMS OF THIS BOOK

The ongoing climate crisis continually unveils our vulnerability as a society. Like peeking behind the curtain, this unveiling means we often see what we might have been happier never knowing about. It is therefore no surprise that many of us choose to avoid news and information on the disastrous impacts climate change has on our lives and the lives of our loved ones: doing so can be frightening and overwhelming. Those of us who do dare to look often suffer from increased climate fear and anxiety. Indeed, these days, climate anxiety is receiving a great deal of attention in the media, among scientists, and within the climate movement.

This book focuses on one flip side of climate anxiety: the courage to practice collective climate action. With this work, we're not aiming to counter anxiety. Indeed, anxiety can be a driving factor behind why people choose to take climate action.[1] What we are aiming to do is to help readers like you build "climate courage" to face this anxiety and avoid falling prey to apathy.

In this book, "climate courage" refers to working resolutely for socio-ecological change in the time of climate crisis, despite climate fears and anxiety. Climate courage as we define it therefore comprises standing up for your convictions and letting yourself feel hopeful and effective. The method of practicing climate courage this book focuses most on is joining small and large groups in order to combat climate injustice. To foster this specific type of courage, it may be beneficial to look at group-based climate action through the lens of psychology. That's where we come in. In this book, the Author Team, comprised of social and environmental psychologists, has summarized state-of-the-art research on collective climate action. We've also shared stories from our own and others' experiences to bridge the gap between research and practice.

Our goal in structuring this book was to make it user-friendly so readers can work with it, learn from it, and implement its insights. While this book consistently points out that there is rarely a one-size-fits-all solution, the recommendations we've provided can be taken as advice, guidelines, or simply inspirations for climate action at the individual or group level. As the Author Team of this book, we believe most people out there would be willing to act with courage in the face of the climate crisis if they only had the tools and resources to help them do so. It is our goal to provide just that.

WHO THIS BOOK IS FOR

This book will be helpful for numerous individuals, particularly those active in the movement for a socio-ecological transformation who are seeking methods for becoming more psychologically resilient as well as effective in their actions. This includes:

	the student organizing weekly protests at her school
	the climate activist blockading a mine in the name of immediate coal phase out
	the critic feeling dissatisfied with individualized solutions to structural problems
You!	the manufacturer trying to make his family business more sustainable
	the ministry worker striving to pass environmental laws in their town
	the agriculture group exploring a new reality of collective socio-ecological practice
	those wanting to better understand the psychology of collective climate action will also find this book of interest

Given the reality of the climate situation we are facing today, those who are courageous enough to act, to seek out joint efforts, and to work towards socio-ecological change deserve our gratitude. It is for these individuals, and those still finding their courage, that this book is written.

THE STORY BEHIND THIS BOOK

The climate crisis is caused by human behavior.[2,3] If we want to prevent the further collapse of the ecosystems that enable life on this planet, we need to change.

In 2016, members of this Author Team published a book on the psychology of environmental protection, with an emphasis on interventions to change individual behavior.[4] While that book garnered substantial attention and helped climate groups improve their campaigns, there was always the concern that focusing on behavior change seemed to spotlight individual responsibility. This is partially due to the topic itself and to the methods used in the field of psychology. Academic psychologists typically investigate the thoughts, feelings, and actions *of individuals* through questionnaires and interviews. It is this very focus of investigation that can potentially be instrumentalized to argue against structural change. In other words, academic psychologists usually investigate the individual, not the collective. Through this lens, it might be easy to say, "if the individual can change, it is their responsibility to do so".

However, researchers in this field consistently state:

> While there are established strategies for fostering behavior change in individuals, it's extremely difficult to achieve profound and lasting change if the societal and physical structures surrounding those individuals do not change.

Psychology-based interventions can be effective but, on average, interventions only lead to a small change.[5] Behaving sustainably within unsustainable structures will not produce the changes we need to see in order to face the climate crisis. What we need is systemic change. This includes changing how our economy works, how we generate and use energy, how we move from one place to another, how we work, and how we spend our time when we're not at work. These changes involve not only our relationship with nature but also our relationships with one another.

Fighting the climate crisis means altering the structure of our society, and this is only possible by working collectively. Confronting the climate crisis comes down not only to changing our individual behaviors, but also to changes on the collective level. Socio-ecological transformation means changes in our socio-political structures and in the minds, hearts, and deeds of the people living within these structures.

For these reasons, and ever since publishing *Psychology of Environmental Protection*, this Author Team has wanted to write a second book that sheds light on the collective side of climate action, focusing on structural change, protest, and collective climate action beyond the private sphere. The world is also a different place than it was in 2016: while the COVID-19 pandemic happened and the war in Ukraine began, we also saw a rise in collective climate action. Within this rise, we're now consistently seeing a lot of people take that first step to engage in climate action, but then step back due to disappointment and burnout. Moreover, the necessary measures for climate protection at the structural level have not been taken, making it highly unlikely that the rise in global temperature will be limited to 1.5°C, opening us up to dangerous warming scenarios of 2°C and above.[6,7] Thus, the urgency of the climate crisis is growing. It now seems more important than ever to build a resilient climate movement that can withstand setbacks and defeat. Climate groups want to and need to prepare for a long-term struggle.

 Box P.1: Food for thought – The trail of climate action

There are those who say, "climate action is not a sprint but a marathon". Our Author Team wonders if this is an understatement – real, enduring climate action seems more like a lifelong hike. And on this trail of climate action, hikers constantly discover new terrain and face new challenges. In this vein, this book can act as a hiking guide, offering ideas on what to bring, what to look out for, what to reflect on, and how we can best reach our destination, together.

DEFINING "WE"

The question of who we are is a fundamental one. The concept of "we" and the role of groups will be explored in greater depth throughout this book, but for now what might be more relevant for you is to know who "we" represents within this text. Since this book is written from the practical perspective of the movement for a socio-ecological transition and thus promotes a better world for all, now and in the future, "we" is often used in this book to denote humanity at large – all of us living here on this planet. Other times, "we" is used to refer to the authors of this book, a.k.a., our Author Team. Over the last several years, these authors have worked collaboratively on this book as part of the NGO *Wandelwerk*.

A QUICK BIT ABOUT *WANDELWERK*

The main goal of *Wandelwerk* is to bring psychology-based insight and support into the environmental and climate movement in order to promote a socio-ecological transformation.

We want to close the gap between science and practice.

As a collective, we are also part of the climate justice movement and have experienced the struggles entailed therewith. In addition to our practical experience, we have scientific backgrounds in environmental and social psychology. This means we are continually reflecting on and studying what motivates people to act, alone or together, especially in the context of the climate crisis.

With roots in both the practical and scientific sides of collective climate action, we're in a unique position to be able to offer two perspectives: that of academics, which may seem somewhat observational and top-down, and that of those involved in the movement, which may seem more involving and bottom-up. Our ability to consistently switch between these perspectives is what makes this book different from, for example, psychological textbooks or books focused solely on experiences from within the movement. Since its founding in 2015, *Wandelwerk* members have hosted numerous lectures, workshops, and interviews on the psychology of the climate crisis. Within this scope, topics have ranged from behavior change, climate communication, and degrowth to team building and collective climate action. We try to be where the big questions of the climate movement are.

OUR AUTHOR TEAM

Our Author Team is comprised of *Wandelwerk* members Karen, Paula, Eva, Sophia, Alex, and Julian, as shown in Image P.1.

Karen Hamann wrote the book *Psychology of Environmental Protection*[4] and co-founded *Wandelwerk* in the same year. Her PhD focused on antecedents and consequences of psychological empowerment in the climate crisis. Currently, she is working at the Department of Social Psychology of Leipzig University on the topics of agency interventions and the energy transition. In writing this book, she has drawn on lessons learned from her PhD, her work at Leipzig University, her experience with engagement in the movement for a socio-ecological transformation, and her training in systemic counseling.

Eva Junge is a co-founder of *Wandelwerk* and has been working as an environmental psychologist since 2015. She specializes in climate communication and the

Image P.1: Our Author Team from left to right: Karen, Julian, our colleague Luise, Eva, Alex, Paula, and Sophia.

Photo by Wandelwerk e.V.

psychology of collective action. She also explores different forms of "artivism" by combining circus arts with direct action. Currently, she works as an advisor for socio-ecological NGOs, providing tools to increase the effectiveness of their work for systemic change.

Paula Blumenschein is a researcher in environmental psychology at TU Dortmund University. In the past, she worked as a transdisciplinary sustainability researcher and studied psychology with a focus on intercultural psychology. For more than ten years now, her work for socio-ecological transformation has included involvement in several sustainability and environmental activist groups on the local and international level.

Sophia Dasch has been involved in various environmental and social justice initiatives since 2012 and is currently an active and board-serving member of *Wandelwerk*. Previously, she worked for the project CO_{2ero} as an environmental educator, accompanying schools on their journey to more climate protection. At Leipzig University, she also investigated the psychology of the energy transition (together with Karen Hamann), and her master's thesis explored how the radical side of a movement impacts its public perception and support. Since 2024, Sophia has been working at ConPolicy, which is a research and consulting firm focusing on consumer policy.

Alex Wernke has been active in the climate movement for 10 years – working on the organization, mobilization, and communication of many of Germany's large-scale climate protests. Since finishing his studies in environmental psychology at the University of Magdeburg, Alex has worked as a facilitator of the participation process Bonn4Future and is currently a freelance political educator and climate communication trainer.

Julian Bleh is a social psychologist currently doing his PhD at Leipzig University. His work focuses on the ability to imagine social change and the motivational effects of having a vision. He is also working as an applied social scientist, facilitating and evaluating projects in the context of socio-ecological change within Germany and is involved in the climate justice and fair housing movements.

Writing this book was part of the *Erasmus+* funded project, *Education for Pro-Environmental Active Citizenship* (EPEAC), in cooperation with the organizations *Ulex* in Spain, *Transformative Education* in the United Kingdom, and *Vedegylet* in Hungary.

 Box P.2: Note – Our position

We must acknowledge the fact that we're writing this book from a very privileged position. All members of our Author Team identify as Caucasian and German and have a university degree. We are not representative of the global climate movement, and readers should keep in mind that the perspective presented in this book is only one of the many valuable perspectives in the field of collective climate action. Due to our backgrounds, examples herein disproportionately stem from the climate movement in Germany. We always welcome hearing from others willing to share their own experiences so that we may tell more comprehensive stories, together. If you have a perspective you'd like to share with us, see the *Final words* section.

TWO CORE QUESTIONS

We, the Author Team, see climate courage in those who act for socio-ecological change despite experiencing fear and doubt. Courage arises in times of anxiety. But how do people keep their spirits up in the face of growing crises? And how can someone muster up the courage to take that first step and join forces with others? Moreover, what motivates us to resist the easier path of apathy? These thoughts can be boiled down to the following two core questions, which this book will focus on:

1. How can we build and sustain collective climate action?

2. How can collective climate action become resilient and effective?

We believe it is essential to learn about people's motivations and perceptions to create impactful socio-ecological change. Throughout this book, we cover a range of scientific topics that are important for gaining such a fundamental understanding. We also aim to provide some practical recommendations that can be directly applied by those who organize and participate in climate actions.

BUILDING A BRIDGE BETWEEN SCIENCE AND PRACTICE

From a scientific standpoint, the Author Team wants to be explicit that, within the field of psychology, research into collective climate action is still in its early stages. The research that does exist, however, builds on a long tradition of psychological scholarship on collective action in other injustice contexts. Although we often wish to provide more specific and detailed solutions for the climate crisis, this is simply not possible if we want such solutions to be thoroughly grounded in empirical findings.

For the past four years, research into collective climate action has gathered momentum. It stands to reason then, that had we waited an additional four years to publish this book, we would have been able to provide even more climate crisis-specific insight. But the simple fact is, the climate crisis is happening *now*. And that means people cannot wait another four years for this knowledge. To quote an activist sticker:

the climate crisis doesn't wait for your bachelor's degree.

It also doesn't wait for researchers to draw final conclusions, were that even possible.

To deal with the limited body of available research when writing this book, we made use of various forms of knowledge: data-driven quantitative knowledge about collective climate action, knowledge about similar topics, knowledge from interview studies, and practical knowledge from our own and others' experiences with collective climate action.

More precisely, we started off by including every psychological study we could find that was practically relevant to the core questions of this book. At the time of writing this book, there were no published scientific reviews or books on the psychology of climate protests and volunteering. So, we conducted our own review, employing a snowball approach, which involved using the references or citations of a paper to identify additional papers, and we interpreted our findings through the lens of the climate movement. While attending conferences, our Author Team also noted down the latest findings of unpublished posters and papers. However, as these findings have not yet gone through the peer review process that ensures scientific quality, readers should consider the results of these unpublished studies preliminary.

Where we did not find any available climate-specific research on an important topic, we considered empirical studies that focus on other social issues, for example, women's and LGBTQ+ rights, the peace movement, and trade unions. For many of these fields, there is a much longer tradition of research. While this book focuses on collective climate action, people involved in other relevant social movements can also use this book as an inspiration for their groups and actions. We complemented these quantitative, data-based findings with qualitative interview studies from the field that provide valuable insights into people's experiences.

Finally, we also share our personal knowledge as people involved in collective climate action, as well as experiences of others who were invited to contribute to this book. We believe that by integrating research and joining discussions we gain inspiring new insights that are relevant to socio-ecological transformations. Aside from its practical value, this book aims to encourage students, researchers, and many others to investigate the questions that arise while reading it.

CRITICALLY REFLECTING ON FINDINGS

This book includes knowledge from very different sources. It is crucial for us as scientists that you as the reader have a basis for evaluating the credibility and assumptions made by the research presented here. An important question is therefore: can causal claims be made based on this research?

Box P.3: Info point – What's causality?

As opposed to co-occurrence, which is the idea that two events simply occur, causality is the idea that one event causes another. An example of this can be seen if we take two events, (1) a person felt morally obliged and (2) joined a protest, and connect them through a causal chain to say that (1 → 2) a person felt morally obliged *so* they joined a protest.

The field of psychology adds to conversations about socio-ecological change as it actually tests causality in experimental designs. However, many of the mechanisms are often not tested experimentally. This leads to co-occurrence and causality sometimes being misconstrued in the public perception of research, a mix-up fake news outlets rely heavily on. Throughout this book, our Author Team has always tried to make it explicit where our assumptions come from and which study designs were used. Nevertheless, we invite you to see our words, assumptions, and advice as preliminary and always question them critically. Throughout this book, you might encounter a few terms of scientific methods you're not yet familiar with. If you're interested, you can take a look at the *Appendix: Overview of research designs* section, where we explain various approaches to quantitative research such as experimental, longitudinal, field interventional, cross-sectional, and meta-analytical studies, as well as other types of research included in this book.

It is also important to stress that, unfortunately, most of the research presented in this book is from individuals falling into the so-called *WEIRD* (western, educated, industrialized, rich, democratic) categorization, which is not representative of the global population. Of course, in the case of the climate crisis, since *WEIRD* individuals are responsible for the majority of emissions,[8] it is worthwhile focusing on this group. Cross-cultural research is still quite rare in environmental psychology, which means conclusions drawn in this book cannot be easily transferred to the Global South. People from the Global South involved in the climate movement are, for example, more likely to suffer from repression. This may influence their feelings, perceptions, and actions, and possibly also the relationships between these psychological processes.

WHAT YOU WILL AND WILL NOT FIND IN THIS BOOK

Let's take a closer look at what you will and will not find in this book in Table P.1.

Table P.1: What you will and will not find in this book

What you will find in this book:	What you will not find:
• A solution-focused approach that intends to work out how people become and stay motivated for collective climate action and contribute to actual change	• A thorough analysis of structural problems around climate change and climate injustice
• Psychology, psychology, psychology … maybe some human ecology	• Many interdisciplinary perspectives, although people from other disciplinary backgrounds are invited to join the discussion around the questions raised
• Hands-on practical advice drawn from previous and ongoing scientific findings and debates	• A scientific debate about what collective climate action is
• Guidance on collective climate action, such as protesting, volunteering, engaging in new socio-ecological practices while also considering the influence of climate action on people outside the movement	• Guidance on individual behavior change, such as dietary choices or sustainable consumption → this is covered in *Psychology for Environmental Protection*, which can be downloaded from our website[a]
• Scientific evidence	• Esoteric assumptions
• Insights from studies conducted in Western, democratic, and neoliberal countries, oftentimes with student samples	• Insights that are universally applicable to all struggles for a socio-ecological transformation around the world

[a] Hamann, K., Löschinger, D. & Baumann, A. *Psychology of Environmental Protection – Handbook for Encouraging Sustainable Actions.* www.wan del-werk.org/en/materialien (2016).

References

1. Wullenkord, M. C., Tröger, J., Hamann, K. R. S., Loy, L. S. & Reese, G. Anxiety and climate change: A validation of the Climate Anxiety Scale in a German-speaking quota sample and an investigation of psychological correlates. *Clim. Change* 168, 20 (2021). https://doi.org/10.1007/s10584-021-03234-6
2. Steffen, W. *et al.* Planetary boundaries: Guiding human development on a changing planet. *Science* 347, 1259855 (2015). https://doi.org/10.1126/science.1259855

3. Richardson, K. *et al.* Earth beyond six of nine planetary boundaries. *Sci. Adv.* (2023).
4. Hamann, K., Löschinger, D. & Baumann, A. *Psychology of Environmental Protection – Handbook for Encouraging Sustainable Actions.* www.wandel-werk.org/en/materialien (2016).
5. Bergquist, M., Thiel, M., Goldberg, M. H. & van der Linden, S. Field interventions for climate change mitigation behaviors: A second-order meta-analysis. *Proc. Natl. Acad. Sci.* 120, e2214851120 (2023). https://doi.org/10.1073/pnas.2214851120
6. Lee, H. *et al.* Synthesis Report of the IPCC Sixth Assessment Report (AR6). (2023).
7. Dooley, K. & Christoff, P. A. Matter of degrees: Why 2C warming is officially unsafe. *The Conversation.* http://theconversation.com/a-matter-of-degrees-why-2c-warming-is-officially-unsafe-42308 (2015).
8. Gore, T. Confronting carbon inequality: Putting climate justice at the heart of the COVID-19 recovery. *Oxfam Media Briefing* (2020). https://oxfamilibrary.openrepository.com/bitstream/handle/10546/621052/mb-confronting-carbon-inequality-210920-en.pdf

ACKNOWLEDGMENTS

This book has been produced as part of the project *Education for Pro-Environmental Active Citizenship* (EPEAC), under the auspices of the European Union program *Erasmus+*. The project has received funding from *Erasmus+*, through a *KA2 Strategic Partnership Grant* under agreement no. KA204-851C2AD9. Neither the European Union (EU) nor the European Education and Culture Executive Agency (EACEA), nor any person acting on behalf of these institutions, is responsible for how the following information is used. The views expressed in this publication are the sole responsibility of the authors and do not necessarily reflect the views of the EU or the EACEA.

This publication was supported by the Open Access Publication Fund of Leipzig University. We, the Author Team, would like to thank all those who have invested their time and effort in giving us feedback on this book, in order to make it both scientifically sound and practically relevant.

Special thanks go to: Aline Kelber, Anna Becker, Anna Castiglione, Annalena Becker, Cornelius Dahm, Dr. Eric Shuman, Felix Formanski, Franca Bruder, Gee, Prof. Gerhard Reese, Prof. Helen Landmann, Ilmari Binder, Karsten Valerius, Katharina von der Kaus, Klara Wenzel, Prof. Laura Henn, Marie Heitfeld, Dr. Marlis Wullenkord, Prof. Martijn van Zomeren, Merle Larro, Prof. Michael Schmitt, and Nathalie Niekisch.

We thank Andreas Bauermeister and Angela Müller for designing an earlier version of the figures featured in this book, as well as Jan Peter Dirks for designing the five icons used in the boxes.

Additional thanks go to our wonderfully talented illustrator, Otto Barboni, for contributing the illustrations at the beginning of each chapter and to our amazingly dedicated and competent language editor, Rebekah Olson at Eco Ventures English. You have really taken the book to a new level.

1

OVERVIEW OF THE PSYCHOLOGICAL MODEL

DOI: 10.4324/9781003558439-1

DEFINING COLLECTIVE CLIMATE ACTION

Collective climate action is at the heart of this book. Collective action is a unique type of action in which both the actions and aims of an individual are connected to a group or movement.[1,2] To put it succinctly:

> ### Collective climate action is when individuals act as members of a group with the aim of changing socio-political structures in the face of the climate crisis.

As you can see, in the field of psychology, we are looking at collective climate actions at the level of individual group members – their beliefs, feelings, and behaviors – and not at the group as a whole.

Collective actions can be what researchers in the field of psychology call *normative*. Examples include peaceful demonstrations, conversations with decision makers, petitions, and information campaigns. Generally, normative actions are broadly accepted by society and do not cross societal boundaries, such as rules, laws, or etiquette. In contrast, there are also *non-normative* types of collective actions. These range from civil disobedience (blocking busy streets, damaging property) and socially unaccepted behavior (nude protests) to violent uprisings (hurting others). In this book, you will find that the distinction between normative and non-normative collective action is relevant not only at the political level but also at the psychological level.[2]

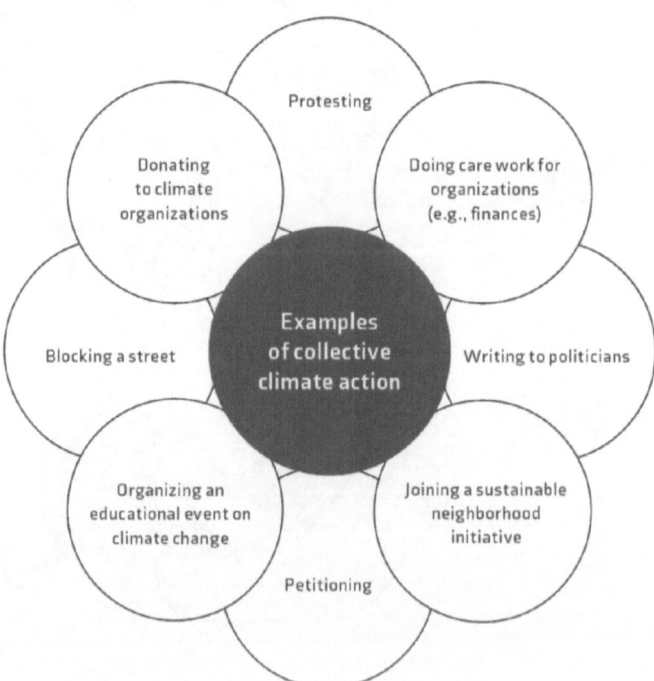

Figure 1.1: Examples of collective climate action

Box 1.1: Note – Activism in this book

Within this book, the Author Team has decided not to speak of activism, as not everyone who joins in collective climate action identifies as an activist. Two exceptions to this can be found in the *Activist's dilemma* and the *Activist burnout* sections, as these employ established terms.

Examples of collective climate action

Collective climate action can take various forms, in addition to those shown in Figure 1.1 and Images 1.1 to 1.3. It can be going door to door to motivate others to join a neighborhood protest. It can be volunteering in an environmental group by joining a plenary session. It can be showing up in public spaces to fight for intergenerational justice. Collective climate action can also be taken by individuals seeking to initiate changes in their respective surroundings, such as citizens writing to their local government officials, employees meeting with their higher-ups, and students talking to their teachers.

As previous psychological research has largely focused on protesting and sometimes on volunteering, these climate actions are most prominent in this book. However, the psychological principles described in relation to these actions are likely transferable to other actions as well.

Image 1.1: Reforestation project in Montana, USA (2020).
Photo by Dave Gardner Creative/ National Forest Foundation (PDM 1.0 Deed)

Image 1.2: Nnimmo Bassey speaks at a camp for climate action in the opencast lignite region in North Rhine-Westphalia, Germany (2016).
Photo by Lars Jung (CC BY 2.0)

DIFFERENTIATING BETWEEN COLLECTIVE AND PRIVATE CLIMATE ACTION

What makes the described examples of action collective is that they do not remain in the private sphere but go beyond the adaptation of personal behaviors to socio-ecological trends. The distinction between collective climate action and private behavior can perhaps be best illustrated through the concepts of the eco-logical footprint and the ecological handprint.[3]

The ecological footprint represents the *negative* impact an individual has on the planet, influenced by their circumstances and lifestyle. Taken seriously, our ecological footprint can weigh heavily on our shoulders. Those of us here in Germany, for example, might feel burdened knowing our individual CO_2 emissions amount to 10.5 tons annually on average – a number significantly higher than the 1 ton that would mean each person living on this planet could have an equal share without us collectively surpassing the global maximum.[4]

An alternative and more solution-focused concept is that of the socio-ecological handprint.[5] The handprint represents the *positive* impact we can have on this planet by creating societal structures that make sustainable behavior easier, more accessible, cheaper, and the new norm. While reducing our footprint might mean eating more sustainably or taking the bike instead of the car, redu-cing our handprint might mean advocating for sustainable food in our school cafeteria or for improved biking infrastructure in our city. While the footprint

Image 1.3: Petition against budget cuts detrimental to the implementation of already planned new bicycle paths in Berlin, Germany (2023).

Photo by Chris Grodotzki/ Campact (CC BY-NC 2.0)

concept often starts and stops with the individual, the handprint concept starts with influence from one person to another and makes its way to societal change. The notion of collective action in this book is closely tied to the concept of the ecological handprint and can be better understood using a specific model called the *Social Identity Model of Collective Action*.

THE SOCIAL IDENTITY MODEL OF COLLECTIVE ACTION

As research psychologists, we as the Author Team like to work with theoretical models. Since models are a simplification of the complex reality, they can help us structure and understand the world. Psychology-based models, in turn, help us structure and understand people's beliefs, feelings, and actions.

One such model has provided the basis for this book: the *Social Identity Model of Collective Action* (SIMCA).[6] Martijn van Zomeren, now a professor of social psychology at the University of Groningen, and colleagues developed the SIMCA to explain the psychological conditions under which people join or intend to join collective action in diverse areas of social change. The model is based on a meta-analysis of studies on anti-globalization demonstrations, farmer protests, sexual discrimination, participation in trade unions, the gay rights movement, the fat acceptance movement, industrial protests, and antinuclear activism, among others. The SIMCA was recently replicated and extended,[7] further increasing its robustness and value.

THE STRUCTURE OF THIS BOOK

The SIMCA, as well as its replication, contains three core concepts used in the structuring of this book: social identity, moral beliefs (including moral convictions and perceived injustice), and efficacy beliefs. If groups and people involved in collective climate action broaden their understanding and reflect on these three core concepts, it is the idea of this book that their efforts will become more effective, enduring, humane, authentic, and resilient in the face of crises. Alongside the SIMCA model, this book uses information gathered from research studying people who are involved in collective climate action, curious about it, or observing this type of action. This approach ensures each chapter contains scientific evidence from the fields of climate change, climate justice, and environmental protection.

Part I of this book is closely tied to scientific findings surrounding these three main concepts, and explicitly describes psychological ideas for and studies on motivating collective action and gaining support. Chapter 2 revolves around people's social identification (our perception that we are part of a group and our emotional attachment to this group).[8] It describes how groups can be designed so that they fulfil people's needs and maintain their involvement. This chapter further sheds light on how people define themselves as part of a climate action group, as illustrated, for example, by the protest chant, "They can't stop the climate revolution; we are the climate solution". Chapter 3 focuses on morals. Moral beliefs are characterized by strong convictions about something being right or wrong regardless of the circumstances.[9] This chapter also covers moral emotions, such as feelings of guilt and anger, which might guide people's moral compass. Moral beliefs can be seen in the impassioned protest chant, "What do we want? Climate justice! When do we want it? Now!" Chapter 4 looks in more detail at how the climate crisis can be communicated to the public with the help of framing. Chapter 5 highlights the role of perceived efficacy in the event of success and in the face of defeat. Efficacy beliefs represent people's perceptions and feelings that they and their groups can achieve their collective aims.[10] This is reflected in the shout, "Power to the people, 'cause the people got the power!" Chapter 6 builds a bridge to the psychological consequences of collective climate action by asking two questions: how do people respond to success and failure, and how does collective climate action influence people's private lives? The concepts of Chapters 2–6 are united in a model for explaining engagement in collective climate action that guides this book (Figure 1.2). Chapter 7 pulls the threads together and offers a suggestion on how to use this model to plan collective climate action.

Part II of this book then opens things up by raising the question of how resilient and effective collective climate action might look. Chapter 8 summarizes up-to-date insights into the causes and consequences of activist burnout and draws conclusions for resilient collective action. Chapter 9 outlines four theories of societal change. As this is not the Author Team's scientific area of expertise, this chapter is meant as an inspirational source for reflecting on your own assumptions about societal change. Chapter 10 then wraps up the book by presenting guidelines for (re)evaluating the goals of climate action groups and for choosing suitable psychology-based strategies to promote engagement for a socio-ecological transformation.

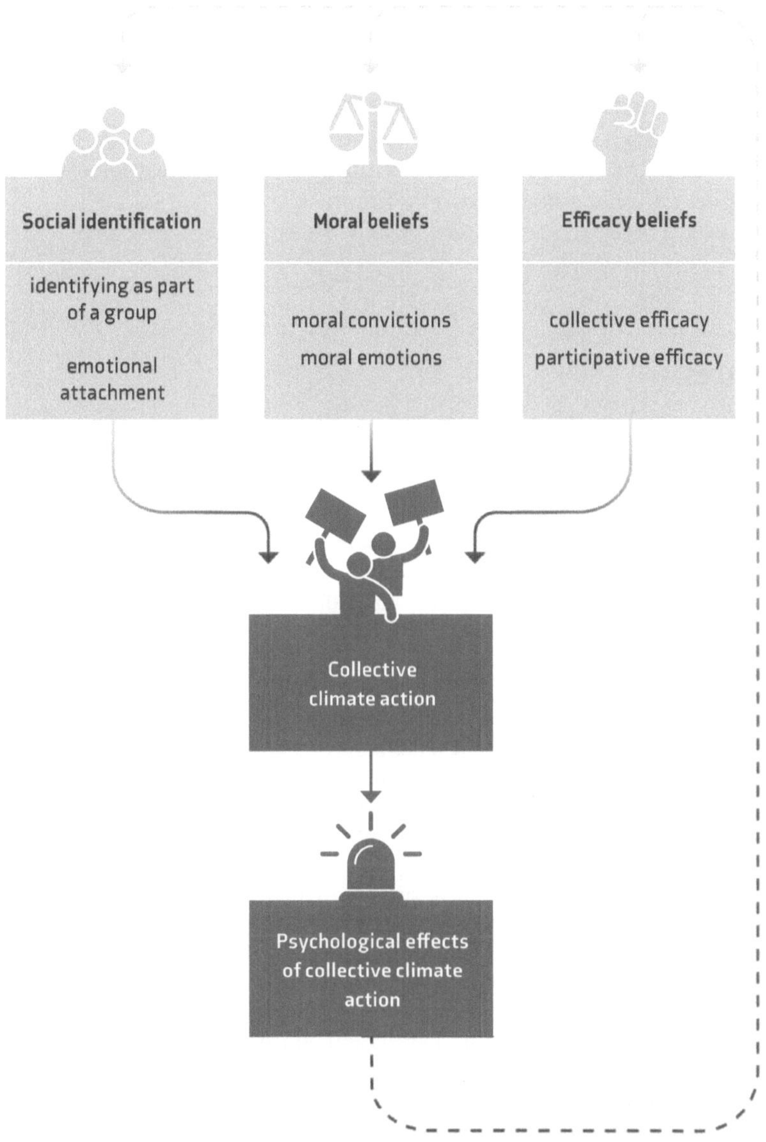

Social identification	Moral beliefs	Efficacy beliefs
identifying as part of a group	moral convictions	collective efficacy
emotional attachment	moral emotions	participative efficacy

Collective climate action

Psychological effects of collective climate action

Figure 1.2: A model for explaining engagement in collective climate action

Box 1.2: Take action – Think about what you want to know

Before you delve into Part I of this book, we invite you to take the opportunity to write down any questions you may have about the psychology of collective climate action. On your phone, in your journal, or on a scrap piece of paper you can store in between the pages of this very book, why not take a moment to jot down a few thoughts – for example, questions pertaining to people's thoughts, feelings, and actions in the context of collective climate action, or your own and others' motivation for collective climate action?

Later in the book, you will be reminded of these questions so that you can reflect on what answers you've found.

References

1. Landmann, H. & Rohmann, A. Being moved by protest: Collective efficacy beliefs and injustice appraisals enhance collective action intentions for forest protection via positive and negative emotions. *J. Environ. Psychol.* 71, 101491 (2020). https://doi.org/10.1016/j.jenvp.2020.101491

2. Wright, S. C., Taylor, D. M. & Moghaddam, F. M. Responding to membership in a disadvantaged group: From acceptance to collective protest. *J. Pers. Soc. Psychol.* 58, 994–1003 (1990). https://doi.org/10.1037/0022-3514.58.6.994

3. Hamann, K. & Masson, T. Kollektives nachhaltiges Handeln und Psychologie. In *Handbuch Globale Kompetenz* (ed. Genkova, P.) 1–16 (Springer Fachmedien Wiesbaden, 2022). https://doi.org/10.1007/978-3-658-30684-7_35-1

4. Bundesumweltministerium. *Kohlenstoffdioxid-Fußabdruck pro Kopf in Deutschland.* https://www.bmuv.de/MD1631 (2023).

5. Reif, A. & Heitfeld, M. *Wandel mit Hand und Fuß. Mit dem Germanwatch Hand Print den Wandel politisch wirksam gestalten: Hintergrundpapier.* https://www.germanwatch.org/de/12040 (2015).

6. van Zomeren, M., Postmes, T. & Spears, R. Toward an integrative social identity model of collective action: A quantitative research synthesis of three socio-psychological perspectives. *Psychol. Bull.* 134, 504–535 (2008). https://doi.org/10.1037/0033-2909.134.4.504

7. Agostini, M. & van Zomeren, M. Toward a comprehensive and potentially cross-cultural model of why people engage in collective action: A quantitative research synthesis of four motivations and structural constraints. *Psychol. Bull.* 147, 667–700 (2021). https://doi.org/10.1037/bul0000256

8. Tajfel, H. Social Categorization, Social Identity and Social Comparison. In *Differentiation between Social Groups: Studies in the Social Psychology of Intergroup Relations* (ed. Tajfel, H.) 61–76 (Academic Press, 1978).

9. Skitka, L. J. & Bauman, C. W. Moral conviction and political engagement: Moral conviction. *Polit. Psychol.* 29, 29–54 (2008). https://doi.org/10.1111/j.1467-9221.2007.00611.x

10. Hamann, K. R. S., Wullenkord, M. C., Reese, G. & van Zomeren, M. Believing that we can change our world for the better: A Triple-A (Agent-Action-Aim) framework of self-efficacy beliefs in the context of collective social and ecological aims. *Personal. Soc. Psychol. Rev.* 28, 11–53 (2024). https://doi.org/10.1177/10888683231178056

Part 1

MOTIVATING COLLECTIVE CLIMATE ACTION

2 SOCIAL **IDENTIFICATION**

DEFINING SOCIAL IDENTIFICATION

On August 20, 2018, Greta Thunberg walked to the Swedish parliament for the first time to strike for the climate. Alone. Just her and her now famous cardboard sign reading, *Skolstrejk för klimatet* [School strike for the climate]. Within a few short months, millions of people had joined her protest. These mostly young

DOI: 10.4324/9781003558439-3

people were suddenly responsible for having formed one of the most influential social movements in recent history, something they had certainly not been planning on just a few months earlier. So, what happened? How did the protest of a 15-year-old girl in Sweden turn into an unprecedented, sustained mass mobilization covering large parts of Europe and the world?

This chapter covers how Greta's protest harnessed the key ingredient for any social movement. Greta made young people aware that they had something in common, that they as a generation will be severely affected by the climate crisis. Upon this realization, they began thinking of themselves as a group – as a "we" – and in the months that followed, they began to act as one. This unification shines through in the *Fridays for Future* chant:

"They can't stop the climate revolution; we are the climate solution."

Such we-feelings and perceptions are the foundation for climate courage. They represent our social identification.

When asked who we are, we typically describe a combination of individual characteristics, such as our name, age, or interests and the characteristics of the groups which we see ourselves as part of, such as gender, occupation, or social movement affiliation. This ability to define ourselves as unique individuals and as members of a group is the basis of one of the most influential theories in psychology, social identity theory.[1]

The basic idea of social identity theory is that the groups we identify with can be understood as an extension of our individual identity: the "I" becomes a "we".[2] As such, social identification comprises two components: the perceptions of the groups we are part of – the box, so to speak, that we put ourselves in – and our emotional connection to these groups.[3] These two aspects may or may not coincide. For example, someone having just moved to the city of London for work might identify as a "Londoner" while not feeling any strong emotional attachment to that identity, given they don't yet know the city or other Londoners very well. Another individual having just moved to the city of London to live near friends might have been dreaming of doing so for a long time and thus feel strongly connected to the "Londoner" identity. Though both of these people can check the box of perceiving themselves as part of the Londoner group, they each have different emotional connections to this group.

There is an infinite number of groups in our lives. These can be ascribed to us by others, derived from what we do in life, and based on shared experiences. They can also be rooted in common opinions.[4] Table 2.1 presents some examples of the features that form the basis of these groups.

All of these group identities influence the way we interpret and experience reality, with good and bad consequences for the climate movement. For example, if a person perceives themselves as having similarities with someone being affected by the climate crisis, they are more likely to act in solidarity with them. On the flip side, if a person perceives themselves as having no similarities with someone, they are more likely to remain distanced.[5] Social identification is

Table 2.1: Examples of features that form the basis of group identities

Features that are ...	Examples
ascribed to us by others	Our age, race, class, gender
derived from our actions	Our occupation, sports, hobbies
based on shared experiences	Realizing we're not alone in being affected and outraged by structural discrimination
rooted in common opinions	Believing that systemic change is needed to confront climate change

at the core of any group that takes the climate crisis seriously and demands systemic change. But social identification also unites those who deny its existence or defend the status quo.[6]

FROM SOCIAL IDENTIFICATION TO COLLECTIVE CLIMATE ACTION

Meta-analyses find that, across contexts and areas of injustice, identification motivates collective action.[7,8] Similarly,

> a wealth of research has shown that whether or not we engage in the fight against the climate crisis is associated with the extent to which we identify with groups that stand for climate action.

Various studies on the *Fridays for Future* protests, *Extinction Rebellion*, an Iranian nature cleaning program, sustainable university initiatives, and the *Transition Town* movement have shown that people who identify more strongly with a climate group are more likely to engage in collective climate action.[9–15]

Now, some of you may be thinking, *I don't identify with a specific group, but I'm still motivated to participate in collective climate action.* This might be because, as studies show, personally identifying as an environmentalist or with environmentalists has also been found to relate to collective climate action.[16,17] Whether our social identification encourages climate action may depend largely on our understanding of what it means to be a member of a particular group – something psychologists call social norms.

Social norms shape our understanding of social groups. They tell us what the members of a given social group do (descriptive norm) and what they think is correct behavior for those within the group, and perhaps those outside it (ought norm or injunctive norm).[18] Social norms provide orientation and direction, so that members of a particular group know how to act and how to think about a given issue.[2] As such, they can be seen as a key process through which social identification is translated into action. For example, if all members of a person's local *Friends of the Earth* group plan to go to the global climate march (descriptive norm), this person may feel an inner obligation to join them. If, during this climate march, the group discusses and agrees that capitalism is at the core of the climate crisis (ought norm), this person might feel inclined to agree with them and then critically reflect on their own ideas around green economy approaches.

Aside from the wealth of research on identification and collective action, research on social norms has mostly focused on private behavior instead of collective action[19]. One study, however, found that people were more willing to participate in a neighborhood initiative for climate protection when they thought their neighbors would participate and expected them to actually do so.[20]

Yet, it is not identifying with just *any* group that brings people into collective action, but specifically with groups that have action-promoting social norms (and that constitute so-called politicized identities). This can be illustrated with an example from another area of collective action: individuals who identified with the *Schwulenverband in Deutschland* [German gay association] were far more likely to join gay rights protests than individuals who identified with the descriptive group, homosexuals[21]. The reason why identification with a social movement is such a powerful basis for collective action may lie within the action-promoting social norms of these groups.[22] The climate movement is an action-oriented group whose social norms revolve around the idea that people should participate in collective climate action (ought norm) and whose members actually do so (descriptive norm). In turn, its members feel an inner obligation to participate.[21,22] Put differently,

> action-promoting social norms are what turn a group that cares about the climate crisis into a group that stands for climate action – into a climate action group.

This is why climate action groups like *Fridays for Future* are especially effective at motivating collective climate action.

Then again, our group identities influence how we perceive and adapt to social norms. As social beings, we are constantly picking up on the action-promoting social norms around us. Whether these norms can indeed influence our actions depends on the attention we give them and our identification with the group promoting them.[18,23] For example, if the *Fridays for Future* movement is important to an individual, they will likely try to join the Friday protests, but if that individual does not identify with the *Fridays for Future* group, the group's actions are less likely to influence that person's decisions. This illustrates the complex ways in which social identification and social norms are intertwined. You may not realize it at first, but when you identify with a group, chances are you'll find yourself thinking and acting similarly to how you perceive fellow group members thinking and acting.

Fortunately, the way in which group identities and social norms influence our climate actions is not set in stone but can be flexible as our perceptions change. Anna Rabinovich and her colleagues from the universities of Exeter, Groningen, and Bath found that when participants were asked to compare their home country (the UK) to another country with greater harmful environmental impact (the US), they perceived the UK as more environmentally friendly and were more willing to take action.[24] In contrast, when they were asked to compare their country to a country with less harmful environmental impact (Sweden), they viewed the UK as less environmentally friendly, and their motivation for action dwindled.

This study shows that even when identifying as members of the same group, individuals' willingness to act on the climate crisis depends on their point of reference. In the next section, we'll explain and showcase many examples of how social identification with climate action groups can be strengthened; for example, through social norms.

 Box 2.1: Food for thought – Models aren't everything

It is worth noting that psychology-based findings speak of averages. On average, people who identify with a group are more likely to join the collective climate actions of that group. However, there might be others for whom this is not the case. For example, a person might build their climate actions around perceived effectiveness, so that each action is rooted in the perceived likelihood of success, without them truly having to feel like part of a group.

While models and findings may help us understand people in general and determine certain rules of thumb, they cannot be automatically transferred to every individual. They can also often explain parts of behavior, small associations, and trends, but not everything. Questions always remain when we delve into what motivates collective climate action. Thus, while models help us understand reality, they never offer a complete representation of it.

HOW WE CAN STRENGTHEN IDENTIFICATION WITH CLIMATE ACTION GROUPS

Whether we identify with a group depends on two social-psychological processes that can promote social identification and, in turn, collective climate action.

First, we need to see ourselves as part of a group (psychologists call this *self-categorization*).[25] This approach (let's call it Focus 1) is about showing people that they are (already) affiliated with a group that stands for climate action, which can be based on commonalities with others. Second, we need to feel emotionally connected to a group and have the motivation to be or become a group member. The second focus (Focus 2) is about creating a group that meets people's needs, so its members are motivated to start and to continue identifying as group members.

In this section, you'll learn about several strategies that build on these two foci.

Focus 1: Highlighting the connection to climate action groups

We identify more strongly with a climate action group when we see commonalities rather than differences between ourselves and other members of the group.

 Box 2.2: Take action – Thought experiment

To get an idea about how identification works, first look up and read a little about the group *Extinction Rebellion (XR)*.
   Then, ask yourself:

- *What similarities do I see between myself and the members of this group?*
- *Is this group's focus on mass extinction in line with how I experience the climate crisis?*
- *Do members of this group look and act like me?*

The more you answered *yes* to these questions, the more likely you identify (at least somewhat) with *XR*. If you found yourself more often answering *no* to these questions, you probably do not identify with this particular climate action group.

Focus 1 – Strategy 1: Identifying common fates and traits

The questions raised bring us to our first strategy for Focus 1. They illustrate two approaches to showing people they have something in common with climate action groups: identifying a common fate and identifying commonalities with group members. These approaches can be used both to strengthen the affiliation to established climate action groups and to encourage the formation of new ones.

Identifying a common fate

Sharing a fate or grievance is often the starting point for people to see themselves as part of the same social movement.[26] And the first step in showing people they have something in common is naturally to draw their attention to that commonality. The idea here is to simply point out to people that they already share fundamental experiences, beliefs, or goals with regard to the climate crisis.

   An example of how this process can be initiated by a group leader can be seen in the formation of the *Fridays for Future* movement. Prior to Greta's appearance on the public stage, there was no shared group identity among young people as a climate action group. Her protest drew public attention to the common fate of young people, namely that their lives will be severely affected by the climate crisis. By doing so, she transformed the loosely defined group *young people* into a clearly delineated, fate-sharing group facing and now fighting the effects of climate change.[27]

Often, a common grievance is connected to an *us-versus-them* conflict. Traditionally, such conflicts are seen as one of the foundations of social identification and collective action.[7] It is the mentality that, in contrast to *them*, we can see who *we* are and what *we* stand for.[28] Greta Thunberg often uses *us-versus-them* messaging in her media appearances when she articulates who her *we* are (younger people sharing the common fate of being affected by the climate crisis) and who *they* are (political and economic elites responsible for the crisis).[27] This mentality can also be illustrated by how some people of the Global South view climate justice. This group might see extreme global injustices (their shared grievance) as structurally disadvantaging their group members (their *us*) while being advantageous for the group members of the privileged Global North (their *them*).

Fostering this kind of *us-versus-them* messaging could be a way of demonstrating that the climate crisis is not just one person's individual fate, but the shared fate of a whole group of people, thus contributing to the development of a group identity. However, there are trade-offs associated with *us-versus-them* messaging. As you will see in some of the other strategies proposed herein, there are a number of reasons why it may also be useful to include our *them* in our *us* when seeking to focus on a joint struggle.

Identifying commonalities with group members

Another important way of determining whether we are part of a group is to compare ourselves with what we think is a typical member of that group.[25] For instance, a person might think of themselves as part of the climate movement because a typical member takes part in climate justice protests and so do they. Another person might see themselves as somewhat affiliated with the climate movement simply because they like a similar style of clothing to what is worn by typical members of that group[29].

So, when communicating with someone who already has a lot in common with a typical member of a given group, it can be helpful to emphasize these commonalities even more. For example, when selecting a group member to welcome newcomers to an organization, it might be beneficial to choose a member who has something in common with the newcomers and not always just a leader or representative, though these positions have their own significance.

Group members in leadership and representative roles are who outsiders look to as defining the typical group member.

The individuals in these roles often represent what a typical group member is or should be. If a leader shares characteristics with the people their group wants to mobilize for climate action, this can form the basis for social identification. In fact, research has shown that what makes a leader motivating is the extent to which they (1) are perceived as being a typical group member, and (2) create a *we-feeling* by demonstrating that they identify with the group.[30]

Furthermore, leaders have the potential to act as role models, showing those with similar characteristics what it means to be a member of their group.[31] For instance, surveys from the climate strike mass demonstrations in Germany in 2019 suggest Greta Thunberg was a role model for other young women. When asked why they came to the demonstrations, young women cited Greta and her impact as particularly influential in their decision to participate, while young men were less likely to do so.[32]

When speaking about the typical group member – the so-called *prototypical member* within the climate movement – it seems relevant to take a look at what this person actually looks like. A review by Sara Vestergren and colleagues from the universities of Linköping and Sussex showed there are indeed certain typical characteristics associated with being an activist for social and/or ecological causes.[33] On average, activists have a higher education, a job in a scientific, social, or creative field, and a lower income; they have fewer marriages and more divorces, fewer children, and more stress in relationships outside their activism.[33] They typically have higher self-esteem, greater wellbeing, and fewer personal worries, but they also report more burnout, possibly due to their choice of work. Participants in *Fridays for Future*, for example, tend to be left-wing and come from households with more books, something that can be seen as an indicator of a person's socio-economic background.[10]

The kind of people that are easily mobilized for the climate movement are likely to have these traits. So, the question is, *do you recognize any of these characteristics in yourself?*

Alongside these characteristics, however, are the stereotypes people hold about environmentalists. Over two qualitative studies, researchers compiled a list of the words Americans most often associated with environmentalists.[34,35] From the positive to the negative, here they are for you:

Cares about the environment, altruistic, self-sacrificing, educated, determined, intelligent, helpful, vegetarian, liberal, drug user, hippie, hairy, unhygienic, unfashionable, over-reactive, eccentric, stubborn, self-righteous, stupid, extreme, aggressive, and militant.

Which of these traits do you identify with this time?

These are "only" stereotypes and thus do not necessarily represent the real world. However, they are important as they determine whether people see commonalities with certain groups or do not want to be associated with them.[36] For example, a person might think they're not part of the group of climate activists because they perceive the typical climate activist as more hardcore than they are. Or maybe they don't want to be associated with this category for fear of what their friends might think. The former was precisely the case found in qualitative interviews of bird watchers, who did not identify with the environmental activist group due to the negative stereotypes associated with that group.[37]

Unpublished work by Robert Gruber and colleagues from the University of Kaiserslautern-Landau further suggests that some people choose to identify less with people who dress in a way they consider typical of climate activists – using multi-colored fabrics, abstract patterns, and coarse textures.[38] Moreover, how climate activists are perceived affects the influence they can have. A study revealed that people were less likely to take private climate action if it was advocated by a prototypical activist.[34] Conversely, a less prototypical environmental activist was less associated with negative stereotypes and thus able to be more convincing.

These findings show that when speaking to an audience that does not engage in protest activity, in addition to highlighting a common fate, it may be useful to portray the climate action group as diverse – to highlight that the group engages in a variety of forms of collective climate action and includes members with varying characteristics. This can be reflected in campaign photos, in the way people talk about their own climate action, or in who is chosen to speak at a protest.

Loosening the typical view people have of environmental activists might make it easier for more people to see themselves engaging in environmental action. Given the abundance of stereotypes of activists and in particular the negative ones, it may be advisable not to refer to group members as climate activists when hoping to involve those who just might believe some of these stereotypes.

In addition to structural causes and agenda setting, perceived similarity is probably also a reason why, in Europe at least, *Fridays for Future* has remained a movement of predominantly White, educated, middle-class people[39]. Other movements have already shown that advocates from marginalized communities can be a powerful force for mobilizing protest participation in their respective communities.[40,41]

Additionally, it may also be a good idea to break down some of the stereotypes of environmentalists in order to make climate action groups more attractive to outsiders. One strategy for doing this is called *positive contact*,[42] which involves talking to people and showing them what members of the climate movement are really like. An example of this is when anti-coal activists in Germany started working closely with those most affected by the industry: the people living in the villages set to be destroyed for lignite mining. Climate activists and people from such rural villages likely come from disparate social groups, have been socialized differently, value different things, and live drastically different lives. While these differences may have contributed to some initial skepticism, it did not take long for most villagers to adjust their views. They probably came to the conclusion that though these activists might be a little strange, more importantly, they're friendly, competent, and hardworking allies in the fight ahead. After a while, people from both groups, the villagers and the activists, worked closely together, and it became less obvious that they came from different walks of life.

Box 2.3: The bottom line

Climate action groups need to let people discover that they have some-thing in common to foster social identification. This can be facilitated by highlighting common fates, presenting a prototypical group member who is relatable to the audience they want to reach, and critically reflecting on and breaking down pre-existing stereotypes. A leader can play a key role in motivating others by presenting as someone who highly identifies with the group and is a prototypical group member.

Focus 1 – Strategy 2: Linking with pre-existing group identities

The fact that identifying with a climate action group like *Fridays for Future* is such a powerful motivation for collective action might lead one to believe that an effective strategy for promoting the fight for climate justice is simply to show people that they belong to such a group. While this may be a promising strategy for people who are already somewhat inclined towards socio-ecological change, large parts of society will not easily begin to see themselves as part of a group that is, in its essence, defined by protecting the climate.

One strategy for dealing with this and mobilizing people for climate action is to link their pre-existing group identities to the fight against the climate crisis, and to show them that the prevailing norms of the groups they're already part of (their neighborhood, their sports club, being a parent) are in line with it. For example, *Parents for Future* focuses specifically on bringing together parents and linking this group identity to climate action. *Fridays for Future* consists of numerous subgroups, such as *Entrepreneurs for Future*, *Scientists for Future*, *Churches for Future*, *Engineers for Future*, *Teachers for Future*, *Grandparents for Future*, *Artists for Future*, and *Psychologists for Future*, to name just a few. Indeed, one of the success factors of *Fridays for Future* may well be that it has been able to form subgroups linked to existing social groups.[43]

There are several approaches to linking pre-existing group identities to the fight against the climate crisis, but for now let's take a look at three of them.

Using value-based communication

One approach for developing the link between pre-existing group identities and the fight against the climate crisis is through value-based communication. This approach involves linking climate action to the goals, beliefs, and values that characterize particular societal groups. Kristin Hurst and Marc Stern from the Department of Forest Resources and Environmental Conservation in Virginia tested this approach by comparing two messages advocating clean energy and a transition away from fossil fuels in the US.[44] Half of their study participants received a message that focused on five universal moral foundations that together are considered relevant to the *conservative* group identity (sanctity, authority, loyalty, care, and fairness), while the other half received a message that focused on only two of these moral foundations and that are typically important to the *liberal* group identity (care, fairness). The message featuring all five morals led to higher support for renewable energy among conservatives than the other message. For liberals, though, both messages were equally effective. These findings indicate that if the aim is to mobilize people from a particular social group, it can under certain conditions be important to tailor messages to their already existing social identities and norms. For conservative target groups, this could mean looking for overlaps between the climate cause and the causes conservatives already value. This way, it might be possible to show people from diverse political backgrounds that tackling the climate crisis is an important part of what they *already* stand for.

If you're looking to better understand what makes people from a particular group tick, there is a range of helpful material out there focused on value-based climate communication. Two of our Author Team's favorites are from *Climate Outreach*: *Britain Talks Climate* and *Übers Klima Reden* [Talking climate][45,46].

👤 Box 2.4: Info point – *Climate Outreach*

Climate Outreach is a British non-profit organization that focuses on public engagement with climate change. Through research, advice, workshops, and training, they help organizations develop new and inspiring climate stories. Feel free to check them out!

Employing block leaders

Another approach for linking pre-existing group identities to the fight against the climate crisis is to employ representatives from target groups. This approach is known as the block leader approach.[6] Research by Tracy Schultz and Kelly Fielding from the University of Queensland showed that community members were more likely to accept recycled drinking water when the information on this issue was provided by a scientist from the same region as community members as opposed to an outsider scientist.[47] This research shows that climate groups may invite representatives from their target group as these are more effective in disseminating climate messaging given their perceived credibility and trustworthiness stemming from a shared identity.[6]

In terms of collective resources, this strategy illustrates that when it comes to trying to convince people of the urgency of the climate crisis,

> it may be more efficient to win over one person from a target group with a strong influence on other group members rather than trying to convince each member of the target group individually.

Identifying an overarching group

Another approach for linking pre-existing group identities to the fight against the climate crisis could be to identify an overarching group[6]. Particularly in contexts where identities are polarized into *us-versus-them* mentalities, this strategy could be useful in building common ground.

In a study illustrating this usefulness, a watershed council was established and included as members stakeholders with varying, sometimes competing, views.[48] Despite their individual differences, their common identity as part of an institutionalized *council* provided the basis for developing joint recommendations for water use in the area. From this, we can take the idea that it may be worthwhile to involve stakeholders from groups that initially appear to be opposed to climate action in processes where they can have a voice and are encouraged to find common ground with others who identify with climate action groups. Good facilitation and institutionalization of these sorts of meetings is key.

The overarching group approach often resonates with people involved in the climate movement because they want everyone to feel affected and be involved, as the climate crisis is a global challenge. These people often highlight identification with all humanity, a special kind of social identification that has been associated with climate action[49]. Indeed, several studies have shown that people who identify more strongly with all humanity are more supportive of the environmental movement and donate more to environmental organizations.[50–52]

 Box 2.5: The bottom line

People's pre-existing group identities can be leveraged to mobilize them for climate action. One way to achieve this is to draw on people's existing groups and show them that what these groups stand for is already in line with climate action. Finding representatives of these groups who are on board with promoting climate action seems to be a particularly promising strategy. There is also the possibility of identifying an overarching group.

 Box 2.6: Note – The debate over how to use group identities

This section has shown that there is no one right way of making use of social identification and social norms but that doing so can help engage people that climate action groups might not otherwise reach. Emphasizing a specific group identity can also have the side effect of reinforcing this identity. In fact, there is an ongoing debate in social psychology as to whether interventions should build on national identification, which is still among the most relevant identities in most countries, or only focus on broader and more inclusive identities. Whatever the strategy, it is important to keep in mind that the messages we put out into the world may be self-perpetuating.

Focus 1 – Strategy 3: Redefining group norms

One way people can find commonalities with climate action groups is to establish action-promoting social norms within their social groups. Whether it's a neighborhood, a university, or a workplace, if we start perceiving action-promoting norms within our group and start seeing other members supporting and engaging in collective action, we are more likely to also see ourselves as part of a climate action group. The strategy here is to actively redefine group norms.

An example of this strategy in action is Greta Thunberg's early public appearances in 2018 and 2019. Through these appearances, Greta established a new action-promoting norm among young people by repeatedly pointing out that "we, the young people, have to act now"[53] (ought norm). She made it clear that young people must put pressure on politicians to make the necessary changes.[54] In doing so, she defined *young people* as a climate action group. At the same time, she participated in the protests every Friday (descriptive norm). Not all young people, of course, but a significant portion of them have since adopted the action-promoting norm that was communicated by Greta.

Here are three of the most effective approaches to communicating social norms.[55]

Highlighting an ought norm

One such approach is to establish an ought norm and define what the group stands for. For example, the *no fly climate sci* community consists of scientists who "feel a need to ramp down unnecessary fossil fuel use" (ought norm) and avoid flying.[56]

Highlighting a descriptive norm

Another option is to highlight a descriptive norm, but only if that norm clearly shows that the majority of the group's members have already taken up action against the climate crisis. For instance, "60% of the people in our community have participated in the climate strike". However, it will often be quite difficult to find these "majority" messages when it comes to collective climate action.

Highlighting a positive trend

The often more appropriate approach to communicating social norms is to highlight a positive trend, such as the increasing number of people taking climate action. This approach is important as some trends might otherwise go unnoticed. An example of this approach at work can be seen in a campaign run in eastern Australia in 2022. Following one of the country's worst recorded floods and a national election in the same year, numerous homes started bearing *CLIMATE ACTION NOW* signs (see Image 2.1). These signs clearly showed the action-promoting norm that there was a majority, or at least a growing number, of people in these neighborhoods who cared about, and would presumably vote for, stricter climate policies. The realization that many of our neighbors (members

Image 2.1: Campaign "Climate Action Now" in Australia (2022).
Photo by Conservation Council ACT Region

of the same group) are in favor of climate action can have a strong impact on our perception of our neighborhood as a climate action group, in the sense that "we as a community stand for climate action".

 Box 2.7: The bottom line

One way people can feel they are part of a climate action group is to establish action-promoting norms within their social groups and highlight how their group has (increasingly) become involved in climate action, whether that group is a neighborhood, local bird watching club, or football team. As norms and norm trends often go unnoticed, highlighting them is a key strategy for changing people's understanding of what it means to be a group member and promoting their motivation for action.

Focus 1 – Strategy 4: Creating a shared group identity

The previously introduced strategies have been largely built on climate action groups or groups not (yet) related to the fight against the climate crisis. Yet, every long-term social movement probably includes the constant formation of new cross-cutting groups with the emergence of new climate-related problems that need solutions. Here, it becomes crucial to build a group identity that is shared.

We need to know how others think about and experience our shared reality in order to arrive at a shared understanding of what *we* as a group stand for.[57] There are two approaches for developing this kind of shared group identity: through interaction and through shared experiences.

Seeking interaction

Historically, social movements such as *Fridays for Future* have often originated in spaces where people interact and engage in constructive dialogue with one another; for example, at universities and schools[58]. Movements also develop at sit-in areas and protest camps, such as New York's Zuccotti Park, which became the staging ground for the *Occupy* movement[59]. Assemblies in Istanbul's Taksim Gezi Park played a role in the formation of the protest movement against worsening authoritarianism in Turkey,[60] and the German anti-coal movement originated in camps, sit-ins, and protest sites in and around forests and villages set to be destroyed for open-cast lignite mining.[61]

Research by Emma Thomas and colleagues of the universities of Murdoch, Western Sydney, and St Andrews offers a clear account of the power of personal interaction in forming new group identities.[62] In their experiment, Thomas and colleagues provided participants with information on the *Water for Life* movement, a movement that aims to provide clean water to people living in the Global South. It was then ensured that all participants supported the movement's mission. After that, half of the participants were instructed to work on their own at developing strategies for promoting this movement. The other half was instructed to do the same but work in small groups of three to five.

It turned out that, compared to doing the task alone, discussing it with others boosted people's identification with other supporters of the movement and, in turn, their commitment to participating in collective action. It seems that finding a shared understanding of how society should change and working on strategies to implement this change is an important step in building identities.[57] Intriguingly, group discussion seems to be especially relevant when one wants to motivate people for more non-normative action, such as civil disobedience.[63]

One takeaway from this study is that in order to promote a shared group identity based on action-promoting norms, it may be beneficial to organize events or create permanent spaces where people can engage in constructive dialogue on climate politics.

> If the formation of a climate action group is perceived as a bottom-up process, this can strengthen people's identification with that group and their motivation for collective climate action.[64]

In the context of the energy transition, bottom-up formation by citizens can even motivate outsiders to join and support energy cooperatives.[65–67] However, bringing people together to strengthen their perception of themselves as a climate action group probably only makes sense if participants are able to engage constructively with each other and, as a result, are able to find common ground on climate action. There are two possible approaches for ensuring this. One is to bring people together who are already somewhat like-minded and able to deal with their differences in a constructive manner; for example, at camps for climate action or movement conferences. The other approach is to ensure that the interaction remains constructive by facilitating the discussion; for example, in facilitated citizens' assemblies.

Having shared experiences

Another way to build and expand a shared or even overarching group identity is to engage in a mutual experience with people from diverging groups. Such an experience can be the catalyst needed for individuals to form a bond with one another, even if they previously saw themselves as part of separate groups.

This is illustrated by the work of John Drury and Steve Reicher from the universities of Sussex and St Andrews, who published several qualitative studies in which people developed a shared group identity by participating in protests together.[68] In one of their studies, two groups, locals and environmental activists, protested against the construction of a road in a UK town.[69] Though the locals and the environmental activists had initially been separated by different levels of involvement, the police didn't discriminate between the two groups, identifying them as one overarching group.

And, vice versa, the overarching group had found a shared opponent in the police. As some of the police actions were perceived as illegitimate, the crowd united around their joint opposition to the police and the authorities behind them. Thus, a new *us-versus-them* mentality emerged. Moreover, being treated as radicals led to the emergence of more radical action-promoting norms within the group, even among those who initially saw themselves as moderates. (For an illustrative example of this, check out Box 2.9.)

Box 2.8: The bottom line

Building and expanding a shared group identity can be supported using two approaches: (1) creating spaces for people to interact constructively and strategize together about how to tackle the climate crisis and (2) bringing different groups together in shared (protest) experiences.

Box 2.9: Food for thought – How do peaceful protests escalate?

An illustrative, non-climate related example of how a peaceful protest escalated is the London *Poll Tax Riots* of 1990.

To analyze how these riots had occurred, researchers Clifford Stott and John Drury from the universities of Abertay and Sussex gathered field notes, video data, participant interviews, written accounts, and even police material and interviews.[70] Here's what they determined had happened.

On March 31, 1990, a group of people organized a peaceful demonstration to protest a controversial taxation system that would have disproportionately affected low-income people. One interviewee said, "A vote was taken as a statement of intent that demonstrators wanted a peaceful march. It looked as though every hand in the park was raised."[70] Although all participants protested the poll tax, at this stage, they were not yet a unified crowd. Though the clear majority engaged in peaceful protest, there was also a confrontational minority of crowd members who threw objects in the direction of the police. Most of the protesters distanced themselves from the confrontational minority.

After a while, some of the demonstrators started a sit-down protest, which was perceived as non-violent and legitimate by fellow protesters. One protester was quoted as saying,

The only slight confrontation was when I got to Downing Street where people had just started doing a sit-down protest. Once every 10 minutes, like, an empty beer can would like clatter and, in fact, reach the pavement sort of 10 feet in front of the police. There was no real threat to public order, or anything like that. [70]

The police, however, viewed the situation differently.

The slight confrontations and the sit-down were indeed perceived as a threat to public order, as well as illegitimate. Moreover, the confrontational minority was viewed as representative of the crowd. This disparate understanding of the situation led the police to intervene. Since they considered the situation dangerous, they started using force to push the protesters towards Trafalgar Square. In doing so, they not only targeted the minority but the crowd as a whole. Said one protester,

They [the police] were just hitting everybody and just being so violent towards everybody. I think the thing that struck me most was that there were people who weren't there for trouble, who were just there for the cause, getting beaten.[70]

It was at this moment that the conflict fully escalated.

The police's use of force changed the dynamic of the crowd. Since the police acted against everyone, their action was perceived as unjust. This collective experience of indiscriminate police brutality unified the protesters. Suddenly, they shared a new group identity and a new shared opponent: the police. One protester recollected, "When we were faced by the police, all the way through people were shouting 'you bastards'. The whole crowd was with each other. Everyone was outraged and were together in a certain feeling and together in what they did."[70]

Moreover, the group that was formerly the confrontational minority became an influential part of the group, leading to shifting social norms. Self-defense against the police was now perceived as legitimate. The

Table 2.2: Typical escalation pattern of a protest

Phase	Protesters	Police
1. Initially peaceful protest	Majority engages in peaceful protest, a confrontational minority exists	Police perceive confrontational minority as representative and the protest as illegitimate
2. Start of conflict	Protesters perceive police action as illegitimate	Police act against the crowd as a whole
3. Protest escalates	Protesters become unified and defense action is perceived as legitimate	Escalation confirms police perceptions and encourages further riot tactics, causing further escalation

unified participants also shared a sense of efficacy among their group, with one such participant saying, "The crowd went forward, because there was so many people and there was quite a strong feeling of power being in such a big group".[70] In turn, the growing resistance of the protesters confirmed the police's previously unjustified perceptions, leading to escalation on both sides. According to one protester:

> The crowd, [...] retaliated throwing sticks, banner poles, bottles – anything they could find. Young people, armed only with placards fought hand to hand with police. Some demonstrators were battened down with truncheons, others had riot shields thrust into their faces.[71]

By the end, the conflict had involved up to 5,000 people and included looting and property destruction by the crowd.

What can we learn from this event? Based on their research, Stott and Drury proposed that two conditions are needed for protests to escalate: first, there must be at least two groups with differing understandings of the situation and its legitimacy; second, there needs to be an unequal distribution of power, such as typically seen between police and protesters.[70] The escalation process is summarized in Table 2.2. However, the researchers also emphasized that this is but one typical pattern of escalation and therefore does not necessarily always emerge.

It may not come as a surprise to seasoned activists that police behavior plays a major role in the escalation of protests. Yet, it remains an important lesson for people who have only limited experience with demonstrations, as participants should be aware of the possibility of encountering violence, even if they don't initiate it.

If a group is planning a protest or any non-normative collective action, it's essential to prepare participants for the fact that situations of conflict may easily escalate under certain conditions.

As seen in this study, such escalation can indeed have the positive effect of altering people's social identification and creating a shared group. However, escalation is probably not the aim of many climate action groups. If a group wants to avoid escalation, it may therefore help to conduct action training, teach de-escalating forms of communication, and publish and advocate a clear action consensus before and during a demonstration.

Focus 2: Creating climate action groups that meet people's needs

Our ability to identify with a group requires more than just perceiving ourselves as affiliated with that group. We also need emotional attachment, combined with motivation to consider ourselves a member of a group. The extent to which being or becoming a member of a group meets our psychological needs determines how motivated we are to identify with that group.[72,73]

Needs are our most fundamental source of motivation; satisfying our needs is what directs our thoughts and actions. Thus, needs are not merely desires but rather "high-quality" motivators. Satisfying our basic psychological needs is crucial for our wellbeing, and when we cannot fulfil them, we feel uncomfortable or stressed.[74,75]

The strategies of Focus 2 will concentrate on four primary needs: the need for belonging, the need for high self-esteem, the need for meaning in life, and the need for control (see Figure 2.1). Within the field of social identity research, these needs have been repeatedly described as part of one framework and empirically tested in their relation to social identification.[74,76–78] The need for belonging is portrayed as overarching because it is considered our most basic social need[74].

The same reasoning applies to the people we may want to motivate to take collective climate action. The more a climate action group meets the needs of

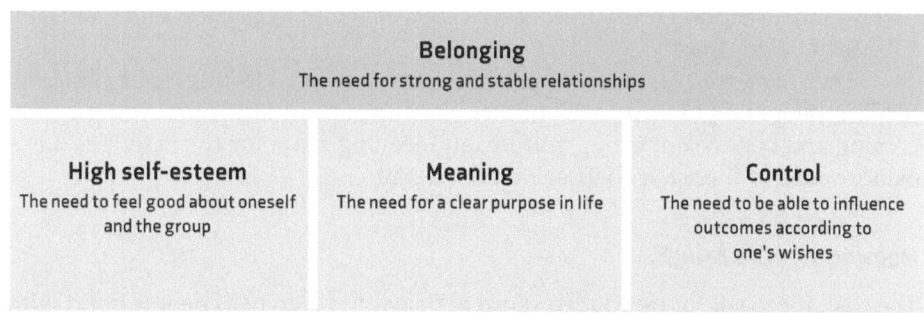

Figure 2.1: Overview of the four basic psychological needs capable of fostering social identification, adapted from Katharine Greenaway and colleagues from the University of Queensland[74]

its members, the more they will identify with it. And, the more a climate action group is perceived by interested non-members as meeting people's needs, the more they will want to identify with it. When writing about the four strategies within this section, we were faced with the fact that research on need satisfaction with respect to climate action groups remains scarce. To compensate for this, we've included many practical examples to illustrate various points.

 Box 2.10: Take action – Thought experiment

If you want to check your own motivation to be part of the climate movement, you can ask yourself, *does the movement …*

give me a sense of belonging?
help me build or maintain social relationships?
make me happy to be part of the movement?
make me feel good about myself?
bring my life meaning?
help me find purpose?
make me feel effective?
give me a sense of control in the face of an otherwise uncontrollable climate crisis?

If your answers to these questions are mostly *yes*, you are probably motivated to identify with the climate movement.

Focus 2 – Strategy 1: Fostering a sense of belonging

Being motivated to belong to groups is the human default.[79] The need for belonging, which is our need to establish and maintain stable relationships with others, is therefore considered our most basic social need.[76]

In the context of collective climate action and environmental action, both published research and our own unpublished studies show that feeling close, related, and connected to other people in one's engagement relates to social iden-tification and taking (protest) action.[80,81] Therefore, if we want people to identify with a climate action group, they need to perceive that this group will give them a sense of belonging.

There are four primary approaches for fostering feelings of belonging: helping people bond with one another, cultivating solidarity, practicing a wel-coming and benevolent group culture, and leaving room for fun activities. Let's explore each of these approaches in more detail.

Helping people bond

The first approach for building a sense of belonging is to help people bond with one another. Research from climate and environmental fields shows that people tend to volunteer more if the activities offer social interaction, if they can meet new people, and if they volunteer with friends and can see familiar faces.[81-84]

One study also found that people identified more strongly with *Fridays for Future* when they perceived that those important to them were also participating in these protests (descriptive norm) and approved of them (ought norm).[10] Thus, climate action groups might be advised to sometimes leave the "efficiency mode" and focus on building real and lasting relationships between their members, in order to sustain their engagement.

This is nicely illustrated in a qualitative study by Anna Reznickova and Lydia Zepeda from the University of Wisconsin-Madison, who interviewed interns and leaders of the *Slow Food* initiative at their university.[85] This initiative organizes slow food events, like family dinners, runs a slow food café, and engages in outreach activity. In one interview, a member vividly describes their feelings of being part of the group and belonging to it:

"My freshman [1st] year was really difficult, and I was having trouble finding people that I really fit in with and had the same interests as me. And it was, I don't know, something just about the people there [at the *Slow Food* initiative]. You sort of felt like you were part of like the big family."[85]

Yet, this feeling may also be threatened when personal contact with other members becomes more difficult. As a matter of fact, the study found that the larger groups become, the more difficult it seems to uphold this feeling of belonging.[85] This could mean that at some stage of the development of a climate action group it might be useful to form smaller subgroups, in which close bonds can still be possible.

A similar problem arose due to the COVID-19 pandemic: many people dropped out of their climate action groups during this time of social distancing and banned group-based gatherings. This may have led to an increased sense of feeling distanced from other group members and fewer opportunities to experience connection with others fighting for the same cause.

What can we learn from this? To motivate identification with climate action groups, we need spaces to forge bonds with other like-minded people. This can be meeting regularly as a local group or less frequently at conferences or camps for climate action, beginning events with a "how are you today?" or ending sessions with a "what are you thankful for today?". At *Wandelwerk*, we use a number of online question generators, which can be found in our reference list.[86–88]

Cultivating solidarity

A second approach to creating a sense of belonging is to cultivate solidarity within climate action groups. Solidarity plays an important role in the formation and cohesion of social groups.[89] To foster people's sense of belonging to climate action groups, groups need to offer a community based on solidary principles.

A straightforward way to promote solidarity is to create permanent solidarity structures within the movement. One example of this can be seen with the German aid organization *Rote Hilfe* [Red Aid], which supports left and progressive activists who find themselves in legal trouble. The organization is funded by subscriptions from its nearly 11,000 members. Over the years, *Rote Hilfe* has covered numerous legal fees, fines, and expenses for people fighting for climate justice in Germany. The organization even supports those who are not members themselves.

Another way to practice solidarity is to support others whose opportunities to participate are limited by marginalization, discrimination, or their daily responsibilities. For example, women and low-income individuals are generally less active in energy communities. However, a report created by some of our research colleagues from the University of Groningen suggests that this is not, as is often assumed, due to a lack of motivation but rather to the fact that members of these two groups often feel unable to join energy communities at all.[65] This may be owing to a lack of spare time, financial resources, or contact with existing members.

This last example shows us that one of the fundamental tasks for climate groups is to ensure that people feel welcome and able to participate regardless of their socio-economic backgrounds. Providing childcare during group activities can enable parents to participate. If people with experiences of structural discrimination, be it on the basis of physical or mental health, sexual orientation, gender identity, ethnicity, or class, are to be part of climate groups, programs must be created that make this possible.

Current research indicates it could be useful for groups to create an empowering self-concept that incorporates people from marginalized groups.[90] This may be done through including a statement on a group's website or through critical reflection on their current practices; for example, whether meeting at a specific location may make people feel excluded due to its connection to academia (universities) or money constraints (cafés, bars).

Something every climate action group should reflect on is whether someone is being presented as their leader, and if so, what characteristics that person has. One study from another collective action area found that in groups led by individuals from high-status backgrounds, such as White and male, individuals from more marginalized backgrounds felt somewhat discouraged from being actively engaged.[91] This is why groups wanting to become more diverse should push for representatives from diverse backgrounds.

> Overall, if we are to better represent our diverse
> societies, we need to make groups more accessible
> to those with less privilege and more demanding
> responsibilities.

Being welcoming

Yet another approach to promoting a sense of belonging in groups is to be respectful, generous, and welcoming (see the welcome sign in Image 2.2).[51,92]

Image 2.2: Land Squat and community garden in Heathrow, London UK (2014).
Photo by Transition Heathrow (CC BY 2.0)

This may sound trivial, but according to a review of Australian volunteer research, negative interactions with others are one of the main reasons people withdraw from volunteer organizations[93]. If people feel that others don't want them there or appear indifferent to their being there, they won't develop a sense of belonging. This is crucial given the fact that left-wing groups are often not perceived as the most welcoming.

In their book *Joyful Militancy*, Carla Bergman and Nick Montgomery cite what they call *rigid radicalism* as the root cause of this problem.[94] According to Bergman and Montogomery, rigid radicalism occurs when members try to conform to perceptions of a radical ideal; for example, in how they speak and behave. Such conforming starts as well-intentioned attempts to create spaces less charged with structural violence. However, rigid enforcement of this conformity can lead people to focus on finding and exposing others' flaws, rather than meeting people where they're at, thereby breeding an insecure and exclusive group culture. Bergman and Montgomery argue that rigid radicalism is one of the reasons why language and behavior in many left-wing social movements are so intensely scrutinized, and why individuals who have not yet learned the rules are often excluded. The researchers note that this reduces people to what they have said or done in particular instances and leads to people being treated as mere symptoms of structural violence, rather than as the complex and ever-changing beings we all are.

It is important to note here that neither Bergman and Mongomery nor we as the Author Team are advocating against naming oppression and enforcing clear boundaries. Beyond this, we suggest there should be a general willingness to engage with differences in the group and to introduce newcomers to the group's ideas and practices in a constructive and generous way. Or as Malcolm X put it, "Don't be in such a hurry to condemn a person because he doesn't do what you do, or think as you think or as fast. There was a time when you didn't know what you know today."[95]

A welcoming culture also means giving newcomers special attention, asking them what they like to do and how they imagine being involved, and including them in activities from the start. Two quotes from the *Slow Food* initiative interviews illustrate this: "I remember going there and being invited to help cook right away".[85] Another participant said, "They give you a space to meet people in a way that's not intimidating. [...] If I'm chopping [food] next to somebody, it's so much easier to engage them in conversation than if I'm sitting next to them on a bus." The second quote mentioned here fits advice from an inspiring TED talk called "Activism needs introverts"[96]. In it, campaigner Sarah Corbett explains that it can be useful to once in a while plan events where potential newcomers who may be more introverted don't have to directly engage in conversations but instead get to do something together, such as sorting vegetables. This way, they can decide whether to talk or listen, and they can feel more comfortable in the organization.

Having fun

A fourth very simple but relevant strategy is fun! In an unpublished study by our Author Team, we found that people involved in the movement for a socio-ecological transformation identified with their climate action group and the overall movement more when they perceived their engagement as fun.[81] This research shows us how important it is for groups to consider what elements they can include in their meetings and events within or outside their climate work that their members would find enjoyable.

At *Wandelwerk*, for example, we try to include lots of *energizers* – casual movement games for activating people's bodies and minds. One of our authors recalls that she once gave a workshop for a group that started to become fascinated with energizers. At first, she did not have that many energizers planned as the group was diverse in age, and their attitude towards energizers was unclear at the onset. Upon noticing people's fascination, she changed her plans and started emphasizing these energizers, using them as tools for getting to know fellow group members. Enhancing feelings of belonging among group members in a fun way, rather than knowledge acquisition, seemed to be the group's strongest need at that point. *Wandelwerk* also regularly organizes short outings to break the monotony of structured meetings. These provide opportunities for fun and rewarding shared experiences. This strategy of weaving fun events into group activities is helpful for creating resilient groups that people feel they want to belong to.

Box 2.11: The bottom line

If we want people to identify with climate action groups, we have to make sure these groups provide a sense of belonging. This involves fostering bond-building among group members through events and friendships, cultivating solidarity (particularly with members from marginalized backgrounds), practicing a respectful, generous culture that is welcoming to newcomers, and actively promoting activities that are fun for all.

Focus 2 – Strategy 2: Making members feel good about the group

Every one of us has a psychological need to maintain a generally positive self-image of ourselves.[76] As depicted in Figure 2.1, we want to feel good about ourselves and have high self-esteem. The need for high self-esteem can also drive our identification with groups as we are motivated to identify with groups that provide a positive view of ourselves and to avoid groups that compromise our self-esteem.[3, 97–101]

Previous research on environmental and climate volunteers has found that the self-esteem drawn from one's volunteering is less important for collective climate action than having other needs met (such as their need for meaning).[81,83,102] Yet people's self-esteem as group members nevertheless seems central. If climate action groups want to enhance their members' self-esteem or if they want to appeal to new members, they can make use of several approaches such as showing them respect and highlighting climate action groups as valued parts of society.

Respecting members and groups

Our self-esteem is influenced by our perception of our social standing and acceptance within our groups. It comes as no surprise that being treated as worthy and with respect by fellow group members increases a member's identification with that group.[103] Members of climate action groups would do well to find ways to explicitly focus on respecting and appreciating each other, such as by sending thank you notes to their members on certain occasions.

Identifying with a group often also means being glad, or even proud, to be a member of that group.[104] In line with this, a wealth of research shows that the perceived social status of a group – the extent to which its members are respected and admired by others[105] – is important for social identification. Climate action groups may therefore increase identification by highlighting that their group is valued within the wider climate movement. As this is true for all climate action groups, respect and appreciation between different groups fighting for the same cause seems crucial – even if their actions are based on different analyses and follow different strategies. Indeed, recent research on the societal impact of social movements suggests the climate movement benefits from having a diverse range of groups using a diverse range of tactics[106] – a good basis for mutual appreciation.

Showing respect and appreciation for each other as different climate action groups does not mean we have to agree with what other groups are doing. In fact, we, the authors, believe it is important to critically assess our various analyses, strategies, and tactics in order to work towards a common goal as a movement. The important point here is,

> ## critique should be formulated in solidarity with other groups with the overarching goal of strengthening the movement as a whole.

What should be avoided at all costs is destructive criticism that ridicules certain groups for doing things differently. Recent examples of such destructive criticism can be seen in some of the reactions within the climate movement to the German climate action group *Letzte Generation* [Last Generation]. Similar to the group *Just Stop Oil*, *Letzte Generation* has engaged in blocking road traffic to fight for radical climate policy. In its early days, many people in the climate movement took issue with the group's tactics. Some questioned whether blocking individual traffic was an effective way to fight for structural change. Others feared that the group's confrontational tactics would backfire. While these are legitimate criticisms, they were often expressed not in a spirit of solidarity, but rather in a mocking way. Being on the receiving end of such destructive criticism is detrimental for the self-esteem of a group's members. Why should someone identify with a climate action group that is seen as naive or reckless?

Such criticism is not rare, and the *Letzte Generation* is not the only climate action group that was recently confronted with it. An excerpt from an email sent around a public distribution list for the German climate movement in August 2022 spoke of the destructive criticism being faced by *Extinction Rebellion (XR)*:

> nevertheless there is always such a bad feeling. *XR* people who come back to the *XR* spaces after action days or after camps for climate action tell of 'FUCK XR' notes in the toilet, people were cut, eyes rolled. After the Justice Now action days last autumn in Berlin, I got feedback from one person that it was perceived as unpleasant that 'so many *XR* people' were running around. As if '*XR* wanted to hijack or take over the whole space'. Of course that was just one person. I found myself ashamed afterwards to see so many familiar *XR* faces at *Justice Now*. How fucked up is that?
> [translated from the original German by the Author Team]

The main point is: a lack of acceptance and being on the receiving end of destructive criticism from those fighting for the same cause crushes the self-esteem of group members. As a consequence, it weakens people's identification with their climate action group, discourages others from joining, and perhaps even weakens other groups within the movement.

Highlighting groups as valued parts of society

At a societal level, people's willingness to identify with climate action groups is likely to depend on the social standing of climate groups. To promote identification, groups could therefore highlight that social movements are a valuable and accepted part of our society. It is often forgotten that, throughout history, social movements have always been crucial for progress, whether it be political, economic, or cultural issues[107–109]. In their time, all of these movements were met with resistance.[110,111] Today, however, most people would agree that it was, for example, worth fighting for better working conditions, for women's right to vote, for the inclusion of disabled people, and for civil rights. Placing the current climate movement in a cohort of other social movements that have improved our societies can help climate action groups better communicate their value.

The societal value of the climate movement can further be highlighted by drawing attention to members who are admired or highly accepted in our societies, such as physicians or scientists, cultural or religious leaders, as well as celebrities from sports or pop culture.[112] There is evidence that these admired people can raise concern about the climate crisis even among groups that are somewhat opposed to climate action, such as conservatives in the US.[113] Whilst their effect on identification with climate action groups has not yet been explored, there is reason to believe that when highly respected people show they are part of a climate action group, their credibility rubs off on that group.

 Box 2.12: The bottom line

One of the foundations of social identification is our psychological need to maintain a reasonable level of self-esteem – something that climate action groups can provide by promoting a culture of respect and appreciation within their own group and between groups of the same movement. It can also be useful to show that climate action groups are a valued part of society, by putting them in a cohort with past respected movements, or by drawing attention to members who are admired in society.

Focus 2 – Strategy 3: Helping to establish clear meaning and purpose

As human beings, we have an innate drive to make sense of ourselves and our (social) environment. We desire clarity about who we are, how we should behave, and what we should think about the world. This is why a highly relevant need is the need for a clear meaning in life,[114] as shown in Figure 2.1. Fulfillment of this need minimizes our own uncertainties, guides and affirms us in our thoughts, feelings, and actions, and helps us identify with groups through which we can find purpose in life.[115]

Indeed, both a study run in the U.S. and a study run in Germany found that individuals involved in the environmental and climate movement are most motivated to volunteer their time for climate action groups when they derive meaning in life from their involvement.[81,102] This particular need for meaning trumped all other needs tested in these studies, and therefore appears to be most essential for collective climate action.

Groups can help us create clear meaning because identifying as a group member provides us with shared meaning, guidance, and affirmation. There are three approaches for doing this: making groups distinguishable, clarifying what a climate action group stands for, and helping to find (new) meaning.

Making groups distinguishable

The first of these approaches, making groups distinguishable, means ensuring a given group is perceived as distinct and group members share relevant characteristics.[116,117] Experimental research has shown that the use of symbols in groups strengthens people's perception that that group has a cohesive, shared meaning.[118] Findings like these point to the importance of having a clear public identity as a climate action group, including a name, a logo, and perhaps even a distinctive look.

> The use of symbols can transform even a diverse and otherwise loosely connected collection of individuals into a distinguishable group.

An example of the power of symbols is the anti-nuclear logo widely known as the "Smiling Sun", which bears the slogan "Nuclear Power? No thanks" and can be seen in Image 2.3. The final version of the logo was designed by Anne

Image 2.3: The Smiling Sun logo.
Image by Anne Lund, SmilingSun-Shop, 1975 (GFDL)

Lund in 1975 in Aarhus, Denmark. Lund wanted a logo that would represent as many people as possible who were against nuclear power, or, as she put it, "a badge that a middle-aged woman would wear on her trench coat".[119] The logo quickly became an iconic representation of the anti-nuclear movement world-wide[120] and is still widely used today. Seeing the logo on people from all walks of life probably helped to make the anti-nuclear group recognizable and to turn it into a meaningful environmental action group.

Clarifying what a group stands for

Clarifying what a climate action group stands for can be another lever for creating shared meaning. Reducing uncertainty through meaningful social identification and norms works best when what defines a group and its members is "simple, clear, unambiguous, focused" and agreed upon.[121,122]

 Box 2.13: Take action – Group project

You might be wondering how your climate action group can come to a clear and agreed-upon identity. Well, one way to find clarity and common ground is to define what you stand for through a participatory process of self-constitution. You can think of this process as like writing and revising an organizational mission statement.[123] It defines the purpose of your group's "us" and gives people a range of options to find their own individual place within this shared identity. The idea here is to put more effort into defining what you stand for as a climate action group, to help members find their own individual clarity about what it means to be a member of the group.

It is important to note that such shared identities should not be mistaken for a call for uniformity. We as humans want to identify with groups but also maintain distinct roles within these groups and to express ourselves freely. This is why people often prefer to identify with groups that welcome this kind of diversity.[124]

One real-life example of successful self-constitution can be seen in the creation of an action consensus for mass civil disobedience within Germany's *Ende Gelände*. *Ende Gelände* is an alliance of many independent climate justice groups, affinity groups, and individual activists. Before any action is taken by these groups, plenary sessions are held to collectively establish common norms in order to form a clear action consensus for all who wish to participate.[125] This action consensus states what "we as *Ende Gelände*" stand for and what participants should and should not do as part of the group actions.

Regarding action taken in protest of coal mines run by the energy company RWE, one such action consensus stated,

we will block and occupy infrastructure that fuels the climate crisis and perpetuates neocolonial relations of exploitation with our bodies. [...]

Our action will present an image of diversity, creativity and openness. Our action is not directed against the workers inside. The safety of all participants is our priority.[125]

Helping to find (new) meaning

Finding and pursuing a purpose in life can lead to the provision of a self-sustaining source of meaning.[126] Having purpose is like having a compass – with a clear goal, our purpose provides us with a sense of direction for how to get there.[127]

In order to promote identification, groups could offer spaces for their members to reflect together on the meaning climate action holds in their lives. For example, members could interview each other about the existential beliefs, values, and goals they pursue in their own work.[128] This may help members to maintain sight of the purpose in their action as group members and, in the long run, sustain identification with their climate action group.

Though not focused on climate action, supplementary research has found that spirituality often provides people with a sense of purpose in life[129]. With that in mind, it may be helpful to have specific rituals in climate action groups as well. Outdoor events, for example, can provide members with the opportunity to reconnect with and explore nature, which also serves to soothe negative emotions[130].

One example of helping people discover their meaning in life through their role in socio-ecological change can be seen in the work of *Iron & Earth*. As a worker-led non-profit organization in Canada, *Iron & Earth*'s mission is to empower workers in the fossil fuel industry in their transition to the renewable energy sector. Their self-presentation video represents a practical example of how to show people from a particular community that they have a purpose in socio-ecological change. It portrays the transformative journey of oil sands workers as they use their valuable skills to embrace renewable energy and become catalysts for change.[131] In the video, viewers explore the workers' life stories through short interviews and witness their adaptation, resilience, and determination. Overall, it shows fossil fuel workers the role they can play and the potential they have to find meaning in the transition to a net zero economy.

 Box 2.14: The bottom line

We are more likely to identify with climate action groups if they can give us a (joint) purpose in life. Therefore, groups should help people find their purpose in the challenges of socio-ecological change and support them in finding their particular role in it. If groups have clear and agreed upon group norms – for example, written down in a vision and mission – and are clearly recognizable – for example, through group symbols – this can be a good foundation for fostering identification.

Focus 2 – Strategy 4: Establishing a sense of control

Lastly, humans have an innate need for control, as shown in Figure 2.1. In order to take a closer look at this need, it can be useful to consider the need for efficacy and the need for autonomy separately, as these have been studied as independent needs in the well-established self-determination theory[132]. We are motivated to identify with groups that give us a sense of self-efficacy within these groups. We also seek a sense of collective efficacy in order to be able to take impactful collective action.[73] This means we want to perceive a clear relationship between what we do and what happens as a result of our actions.[76] Interestingly, when our self-efficacy is threatened, we tend to look for groups that are effective or perceive groups that we already belong to as more effective.[133] Thereby, being part of an influential climate action group can give us a sense of collective control over what happens in the political arena.[134] In fact, the need for efficacy and perceptions around it are so central to collective climate action that an entire chapter of this book has been dedicated to these topics (Chapter 5).

Next to this need for efficacy, people strive for a self-determined life. Group members need to feel that they can autonomously determine their own actions without being told what to do. This is reflected in a wealth of research on need satisfaction.[132] In establishing a group culture, autonomy-supportive language (phrases with "can" or "could" instead of "must" or "should"), non-hierarchical structures, and collective decision making may contribute to satisfying people's need for autonomy.[135]

The following quotes[85] – taken from a leader and an intern, respectively, of the *Slow Food* initiative mentioned in this section's Focus 2 – Strategy 1 – illustrate that feelings of self-determination can be integrated into a group culture and can have a profound impact on that group's members and their experiences:

"The best part about it was working with the interns, and my particular leadership style was, 'you've got an idea? Great, let's run with it.' And really allowing students to kind of go where they wanted and to really take ownership of their internship."

"[The initiative] is kind of unique, where no one told us we could do whatever we wanted, but we kind of just did whatever we wanted. And so we were able to get any kind of experience that was needed and kind of make [...] the best out of it."

Interestingly, these interviews from Reznickova and Zepeda also highlight how the different needs we've described thus far can be intertwined.[85] For example, if we are able to pick our own tasks (need for autonomy), we are likely to choose tasks that we can perform successfully (need for efficacy) and that might give us a sense of purpose (need for meaning in life). Experiences of success can then be amplified by sharing them with other group members (need for belonging), which may make us feel good about ourselves and our group (need for self-esteem). In turn, if we then feel close to other group members and trust them, a group culture can flourish in which members dare to design their own tasks (autonomy). Budding research indicates that the need for efficacy might be especially important during early stages of engagement, while the need for belonging becomes relevant later on.[135,136]

 ## Box 2.15: The bottom line

We want to be able to exert some degree of control over our social surroundings. So, we identify with groups in order to experience self-efficacy and become influential as a collective actor. Within groups, making members feel competent and self-determined in their actions is a key ingredient for identification.

 ## Box 2.16: Note – The responsibility for functioning groups

In the view of the Author Team, the responsibility for functioning groups and communities and the application of the steps described in this section lie not only with the members of climate groups but also with governments, businesses, and other actors. For example, researchers (including those from our Author Team) argue that in order for people to play an active part in the energy transition, it is also governments' responsibility to provide the structures and support to ensure that energy cooperatives can successfully form from the bottom up.[67]

DISCOVERING YOUR SOCIAL IDENTIFICATION

Throughout this book, you'll notice we've included a few Take Action boxes – these tasks are meant to encourage you to reflect on what you're reading and aid in your understanding of it.

If you'd like to expand on these tasks, try out the following exercise. You can use it to get an idea of which people and groups influence you, find out how much you identify with these groups, and investigate how well these groups meet your own needs. For this exercise, we recommend using a physical sheet of paper, or little badges, so that you can easily move them, though you're welcome to try just envisioning one instead.

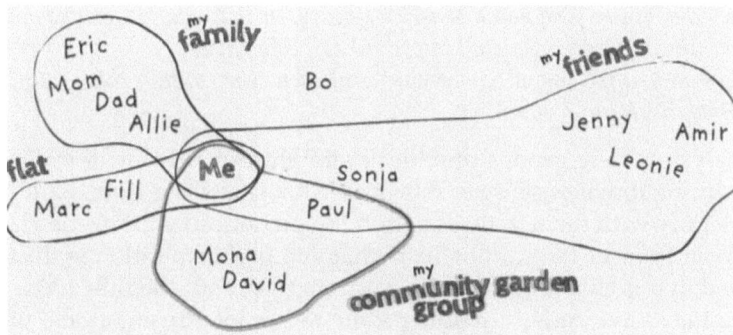

Figure 2.2: Example of Step 1

Step 1: Getting an overview of your social groups and influences

1. Start by writing your own name in the center of the page. If you fancy yourself an artist, feel free to draw a person wherever we say write a name, throughout this exercise.
2. Around your name, write the names of the people who are relevant to you, such as your best friend. The closer you feel to a person, the nearer their name should be to yours. You can also choose people you spend a lot of time with even if you don't necessarily feel close to them, such as your colleagues. Ten names should be sufficient here, but of course you can write down as many as you want.
3. Circle specific groups. For example, you might form one group with your fellow climate action group members and another with your colleagues. From this you will be able to see which and how many groups you are part of. These can be groups that have been ascribed to you by others or the ones that you actually feel like you belong to.

Step 1 should already give you a good overview of which people and groups most likely make up your primary sphere of influence. Figure 2.2 shows a fictional example of Step 1 that was created by a member of our Author Team:

Step 2: Reflecting on your groups
In this second step, make a list of the groups you identified in Step 1. We suggest starting with the one you're most eager to reflect on. Note this may stir up some emotions that you may want to discuss with someone else afterwards. Thinking about this group, do your best to answer the following questions:

- *What are the social norms of this group?*
- *How do people in this group typically act?* (descriptive norm)
- *What do group members think about how people ought to behave?* (ought norm)
- *Does this group meet your needs? Why or why not?*
- *Do you feel you truly belong to and are welcome in this group?*
- *Does this group make you feel good about yourself, raise your self-esteem?*
- *Does this group bring meaning to your life or help you find meaning?*

- *Does this group give you a sense of efficacy, in that you feel competent and self-determined as a member, and impactful as a collective?*
- *What do you like about how people treat each other within this group? What do you not like?*

Analyzing these groups provides a straightforward way of leaving a positive handprint; it makes sense to reflect on what is missing from a group to help you to identify with them in the long run. If you noticed your needs aren't being met, it is our hope as the Author Team that you find inspiration in the strategies mentioned in this chapter of how to (re)strengthen your identification.

You may have ended up doing some rather intense emotional work here. So, take a break, have a cookie, chat with a friend. And when you're ready, we'll see you in the next chapter, which delves into one of the primary reasons we join climate action groups in the first place – our moral beliefs.

References

1. Tajfel, H. Social Categorization, Social Identity and Social Comparison. In *Differentiation between Social Groups: Studies in the Social Psychology of Intergroup Relations* (ed. Tajfel, H.) 61–76 (Academic Press, 1978).
2. Fritsche, I., Barth, M., Jugert, P., Masson, T. & Reese, G. A. Social Identity Model of Pro-Environmental Action (SIMPEA). *Psychol. Rev.* 125, 245–269 (2018). https://doi.org/10.1037/rev0000090
3. Tajfel, H. & Turner, J. C. The Social Identity Theory of Intergroup Behavior. In *Political Psychology* (eds. Jost, J. T. & Sidanius, J.) 276–293 (Psychology Press, 1986).
4. McGarty, C., Bliuc, A.-M., Thomas, E. & Bongiorno, R. Collective action as the material expression of opinion-based group membership. *J. Soc. Issues* 65, 839–857 (2009). https://doi.org/10.1111/j.1540-4560.2009.01627.x
5. Lizzio-Wilson, M., Mirnajafi, Z. & Louis, W. R. Who We Are and Who We Choose to Help (or Not): An Introduction to Social Identity Theory. In *Solidarity and Social Justice in Contemporary Societies: An Interdisciplinary Approach to Understanding Inequalities* (eds. Yerkes, M. A. & Bal, M.) 17–28 (Springer International Publishing, 2022). https://doi.org/10.1007/978-3-030-93795-9_2
6. Fielding, K. S. & Hornsey, M. J. A social identity analysis of climate change and environmental attitudes and behaviors: Insights and opportunities. *Front. Psychol.* 7, 121 (2016). https://doi.org/10.3389/fpsyg.2016.00121
7. van Zomeren, M., Postmes, T. & Spears, R. Toward an integrative social identity model of collective action: A quantitative research synthesis of three socio-psychological perspectives. *Psychol. Bull.* 134, 504–535 (2008). https://doi.org/10.1037/0033-2909.134.4.504
8. Agostini, M. & van Zomeren, M. Toward a comprehensive and potentially cross-cultural model of why people engage in collective action: A quantitative research synthesis of four motivations and structural constraints. *Psychol. Bull.* 147, 667–700 (2021). https://doi.org/10.1037/bul0000256
9. Wallis, H. & Loy, L. S. What drives pro-environmental activism of young people? A survey study on the Fridays For Future movement. *J. Environ. Psychol.* 74, 101581 (2021). https://doi.org/10.1016/j.jenvp.2021.101581
10. Haugestad, C. A. P., Skauge, A. D., Kunst, J. R. & Power, S. A. Why do youth participate in climate activism? A mixed-methods investigation of the #FridaysForFuture

climate protests. *J. Environ. Psychol.* 76, 101647 (2021). https://doi.org/10.1016/j.jenvp.2021.101647

11. Landmann, H. & Naumann, J. Being positively moved by climate protest predicts peaceful collective action. *Glob. Environ. Psychol.* https://www.psycharchives.org/en/item/72fb35c7-7174-47b6-985c-7d1a477965db (2023).

12. Furlong, C. & Vignoles, V. L. Social identification in collective climate activism: Predicting participation in the environmental movement, extinction rebellion. *Identity* 21, 20–35 (2021). https://doi.org/10.1080/15283488.2020.1856664

13. Keshavarzi, S., McGarty, C. & Khajehnoori, B. Testing social identity models of collective action in an Iranian environmental movement. *J. Community Appl. Soc. Psychol.* 31, 452–464 (2021). https://doi.org/10.1002/casp.2523

14. Hamann, K. R. S., Holz, J. R. & Reese, G. Coaching for a sustainability transition: Empowering student-led sustainability initiatives by developing skills, group identification, and efficacy beliefs. *Front. Psychol.* 12, 623972 (2021). https://doi.org/10.3389/fpsyg.2021.623972

15. Bamberg, S., Rees, J. & Seebauer, S. Collective climate action: Determinants of participation intention in community-based pro-environmental initiatives. *J. Environ. Psychol.* 43, 155–165 (2015). https://doi.org/10.1016/j.jenvp.2015.06.006

16. Dono, J., Webb, J. & Richardson, B. The relationship between environmental activism, pro-environmental behaviour and social identity. *J. Environ. Psychol.* 30, 178–186 (2010). https://doi.org/10.1016/j.jenvp.2009.11.006

17. Fielding, K. S., McDonald, R. & Louis, W. R. Theory of planned behaviour, identity and intentions to engage in environmental activism. *J. Environ. Psychol.* 28, 318–326 (2008). https://doi.org/10.1016/j.jenvp.2008.03.003

18. Cialdini, R. B., Reno, R. R. & Kallgren, C. A. A focus theory of normative conduct: Recycling the concept of norms to reduce littering in public places. *J. Pers. Soc. Psychol.* 58, 1015–1026 (1990).

19. Bergquist, M., Nilsson, A. & Schultz, W. P. A meta-analysis of field-experiments using social norms to promote pro-environmental behaviors. *Glob. Environ. Change* 59, 101941 (2019). https://doi.org/10.1016/j.gloenvcha.2019.101941

20. Rees, J. H. & Bamberg, S. Climate protection needs societal change: Determinants of intention to participate in collective climate action: Collective climate action intention. *Eur. J. Soc. Psychol.* 44, 466–473 (2014). https://doi.org/10.1002/ejsp.2032

21. Stürmer, S. & Simon, B. The role of collective identification in social movement participation: A panel study in the context of the German gay movement. *Pers. Soc. Psychol. Bull.* 30, 263–277 (2004). https://doi.org/10.1177/0146167203256690

22. van Zomeren, M. Building a tower of babel? Integrating core motivations and features of social structure into the political psychology of political action: Motivation, structure, and action. *Polit. Psychol.* 37, 87–114 (2016). https://doi.org/10.1111/pops.12322

23. Masson, T. & Fritsche, I. Adherence to climate change-related ingroup norms: Do dimensions of group identification matter?: Adherence to climate change-related ingroup norms. *Eur. J. Soc. Psychol.* 44, 455–465 (2014). https://doi.org/10.1002/ejsp.2036

24. Rabinovich, A., Morton, T. A., Postmes, T. & Verplanken, B. Collective self and individual choice: The effects of inter-group comparative context on environmental values and behaviour. *Br. J. Soc. Psychol.* 51, 551–569 (2012). https://doi.org/10.1111/j.2044-8309.2011.02022.x

25. Turner, J. C., Hogg, M. A., Oakes, P. J., Reicher, S. D. & Wetherell, M. S. *Rediscovering the Social Group: A Self-Categorization Theory* (Basil Blackwell, 1987).

26. Simon, B. & Klandermans, B. Politicized collective identity: A social psychological analysis. *Am. Psychol.* 56, 319–331 (2001). https://doi.org/10.1037/0003-066X.56.4.319

27. Bleh, J. What do we want!? Identität, Moral und Wirksamkeit: Eine sozialpsychologische Perspektive auf die Erfolgsfaktoren der jungen Klimabewegung. In *Climate Action – Psychologie der Klimakrise: Handlungshemmnisse und Handlungsmöglichkeiten* (eds. Dohm, L., Peter, F. & van Bronswijk, K.) 251–282 (Psychosozial-Verlag, 2021). https://doi.org/10.30820/9783837978018-251

28. van Zomeren, M., Kutlaca, M. & Turner-Zwinkels, F. Integrating who "we" are with what "we" (will not) stand for: A further extension of the Social Identity Model of Collective Action. *Eur. Rev. Soc. Psychol.* 29, 122–160 (2018). https://doi.org/10.1080/10463283.2018.1479347

29. Hester, N. & Hehman, E. Dress is a fundamental component of person perception. *Personal. Soc. Psychol. Rev.* 108886832311579 (2023). https://doi.org/10.1177/10888683231157961

30. Steffens, N. K., Schuh, S. C., Haslam, S. A., Pérez, A. & van Dick, R. 'Of the group' and 'for the group': How followership is shaped by leaders' prototypicality and group identification. *Eur. J. Soc. Psychol.* 45, 180–190 (2015). https://doi.org/10.1002/ejsp.2088

31. Khumalo, N., Dumont, K. B. & Waldzus, S. Leaders' influence on collective action: An identity leadership perspective. *Leadersh. Q.* 33, 101609 (2022). https://doi.org/10.1016/j.leaqua.2022.101609

32. Sommer, M., Haunss, S., Gardner, B. G., Neuber, M. & Rucht, D. Wer demonstriert da? Ergebnisse von Befragungen bei Großprotesten on Fridays for Future in Deutschland im März und November 2019. In *Fridays for Future – Die Jugend gegen den Klimawandel* (eds. Sommer, M. & Haunss, S.) 15–66 (transcript, 2020). https://doi.org/10.1515/9783839453476-002

33. Vestergren, S., Drury, J. & Chiriac, E. H. The biographical consequences of protest and activism: A systematic review and a new typology. *Soc. Mov. Stud.* 16, 203–221 (2017). https://doi.org/10.1080/14742837.2016.1252665

34. Bashir, N. Y., Lockwood, P., Chasteen, A. L., Nadolny, D. & Noyes, I. The ironic impact of activists: Negative stereotypes reduce social change influence. *Eur. J. Soc. Psychol.* 43, 614–626 (2013). https://doi.org/10.1002/ejsp.1983

35. Klas, A., Zinkiewicz, L., Zhou, J. & Clarke, E. J. R. "Not all environmentalists are like that … ": Unpacking the negative and positive beliefs and perceptions of environmentalists. *Environ. Commun.* 13, 879–893 (2019). https://doi.org/10.1080/17524032.2018.1488755

36. Stuart, A., Thomas, E. F. & Donaghue, N. "I don't really want to be associated with the self-righteous left extreme": Disincentives to participation in collective action. *J. Soc. Polit. Psychol.* 6, 242–270 (2018). https://doi.org/10.5964/jspp.v6i1.567

37. Cherry, E. "Not an environmentalist": Strategic centrism, cultural stereotypes, and disidentification. *Sociol. Perspect.* 62, 755–772 (2019). https://doi.org/10.1177/0731121419859297

38. Gruber, R., Kachel, S., Menke, S. & Loy, L. S. Vestimentary Manifestations of Group Stereotypes. (presented at 19th Meeting of the EASP, 2023).

39. Neuber, M., Kocyba, P. & Gardner, B. The Same Only Different: Die Fridays for Future-Demonstrierenden im europäischen Vergleich. In *Fridays for Future – Die Jugend gegen den Klimawandel* (eds. Sommer, M. & Haunss, S.) 67–94 (transcript, 2020). https://doi.org/10.14361/9783839453476-003

40. Harlow, S. & Benbrook, A. How #Blacklivesmatter: Exploring the role of hip-hop celebrities in constructing racial identity on Black Twitter. *Inf. Commun. Soc.* 22, 352–368 (2019). https://doi.org/10.1080/1369118X.2017.1386705

41. Towler, C. C., Crawford, N. N. & Bennett, R. A. Shut up and play: Black athletes, protest politics, and black political action. *Perspect. Polit.* 18, 111–127 (2020). https://doi.org/10.1017/S1537592719002597

42. Pettigrew, T. F. & Tropp, L. R. A meta-analytic test of intergroup contact theory. *J. Pers. Soc. Psychol.* 90, 751–783 (2006). https://doi.org/10.1037/0022-3514.90.5.751

43. Together for Future e.V. *For Future Bündnis.* https://www.for-future-buendnis.de/

44. Hurst, K. & Stern, M. J. Messaging for environmental action: The role of moral framing and message source. *J. Environ. Psychol.* 68, 101394 (2020). https://doi.org/10.1016/j.jenvp.2020.101394

45. Wang, S., Corner, A. & Nicholls, J. *Britain Talks Climate: A Toolkit for Engaging the British Public on Climate Change.* (Climate Outreach, 2020). http://doi.org/10.13140/RG.2.2.17707.67362

46. Melloh, L., Rawlins, J. & Sippel, M. *Übers Klima reden: Wie Deutschland beim Klimaschutz tickt. Wegweiser für den Dialog in einer vielfältigen Gesellschaft.* (Climate Outreach, 2022).

47. Schultz, T. & Fielding, K. The common in-group identity model enhances communication about recycled water. *J. Environ. Psychol.* 40, 296–305 (2014). https://doi.org/10.1016/j.jenvp.2014.07.006

48. Samuelson, C. D., Peterson, T. R. & Putnam, L. Group Identity and Stakeholder Conflict in Water Resource Management. In *Identity and the Natural Environment* (eds. Clayton, S. & Opotow, S.) 273–296 (The MIT Press, 2003). https://doi.org/10.7551/mitpress/3644.003.0017

49. McFarland, S. *et al.* Global human identification and citizenship: A review of psychological studies. *Polit. Psychol.* 40, 141–171 (2019). https://doi.org/10.1111/pops.12572

50. Leung, A. K.-Y., Koh, K. & Tam, K.-P. Being environmentally responsible: Cosmopolitan orientation predicts pro-environmental behaviors. *J. Environ. Psychol.* 43, 79–94 (2015). https://doi.org/10.1016/j.jenvp.2015.05.011

51. Renger, D. & Reese, G. From equality-based respect to environmental activism: Antecedents and consequences of global identity: Respect and global identity. *Polit. Psychol.* 38, 867–879 (2017). https://doi.org/10.1111/pops.12382

52. Pong, V. & Tam, K.-P. Relationship between global identity and pro-environmental behavior and environmental concern: A systematic review. *Front. Psychol.* 14, 033564 (2023). https://doi.org/10.3389/fpsyg.2023.1033564

53. Thunberg, G. & Taylor, A. Think we should be at school? Today's climate strike is the biggest lesson of all. *The Guardian.* https://www.theguardian.com/commentisfree/2019/mar/15/school-climate-strike-greta-thunberg (2019).

54. Sommer, M. & Haunss, S. Fridays for Future: Eine Erfolgsgeschichte vor neuen Herausforderungen. In *Fridays for Future – Die Jugend gegen den Klimawandel* (eds. Sommer, M. & Haunss, S.) 237–252 (transcript, 2020). https://doi.org/10.1515/9783839453476

55. Hamann, K., Löschinger, D. & Baumann, A. *Psychology of Environmental Protection – Handbook for Encouraging Sustainable Actions.* www.wandel-werk.org/en/materialien (2016).

56. No Fly Climate Sci. Welcome. https://noflyclimatesci.org/

57. Smith, L. G. E., Thomas, E. F. & McGarty, C. "We must be the change we want to see in the world": Integrating norms and identities through social interaction: The identity-norm nexus. *Polit. Psychol.* 36, 543–557 (2015). https://doi.org/10.1111/pops.12180

58. Teune, S. Schulstreik – Geschichte einer Aktionsform und die Debatte über zivilen Ungehorsam. In *Fridays for Future – Die Jugend gegen den Klimawandel* (eds. Sommer, M. & Haunss, S.) 131–146 (transcript, 2020). http://doi.org/10.14361/9783839453476-006

59. Graeber, D. *The Democracy Project: A History, a Crisis, a Movement.* (Spiegel & Grau, 2013).

60. Uysal, M. S. & Akfırat, S. A. Formation of an emergent protestor identity: Applying the EMSICA to the Gezi Park protests. *Group Process. Intergroup Relat.* 25, 527–539 (2022). https://doi.org/10.1177/1368430220983597

61. Mohr, A. & Smits, M. Sense of place in transitions: How the Hambach Forest Movement shaped the German coal phase-out. *Energy Res. Soc. Sci.* 87, 102479 (2022). https://doi.org/10.1016/j.erss.2021.102479

62. Thomas, E. F., McGarty, C. & Mavor, K. Group interaction as the crucible of social identity formation: A glimpse at the foundations of social identities for collective action. *Group Process. Intergroup Relat.* 19, 137–151 (2016). https://doi.org/10.1177/1368430215612217

63. Thomas, E. F., McGarty, C. & Louis, W. Social interaction and psychological pathways to political engagement and extremism: Pathways to political engagement and extremism. *Eur. J. Soc. Psychol.* 44, 15–22 (2014). https://doi.org/10.1002/ejsp.1988

64. Jans, L. Changing environmental behaviour from the bottom up: The formation of pro-environmental social identities. *J. Environ. Psychol.* 73, 101531 (2021). https://doi.org/10.1016/j.jenvp.2020.101531

65. Goedkoop, F., Jans, L., Perlaviciute, G. & Held, J. *Report on experimental studies on energy communities – Effects of energy community set-ups on support for and willingness to join energy communities.* https://ec2project.eu/resources/downloads (2023).

66. Hamann, K., Masson, T., Fritsche, I., Dasch, S. & von der Kaus, S. *Report on experimental lab studies on energy citizenship – Energy community set-ups, energy visions and collective agency as predictors of energy citizenship and pro-environmental spillover.* https://ec2project.eu/resources/downloads (2023).

67. Hamann, K. R. S. *et al.* An interdisciplinary understanding of energy citizenship: Integrating psychological, legal, and economic perspectives on a citizen-centred sustainable energy transition. *Energy Res. Soc. Sci.* 97, 102959 (2023). https://doi.org/10.1016/j.erss.2023.102959

68. Drury, J. & Reicher, S. Collective psychological empowerment as a model of social change: Researching crowds and power. *J. Soc. Issues* 65, 707–725 (2009). https://doi.org/10.1111/j.1540-4560.2009.01622.x

69. Drury, J., Reicher, S. & Stott, C. Transforming the boundaries of collective identity: From the 'local' anti-road campaign to 'global' resistance? *Soc. Mov. Stud.* 2, 191–212 (2003). https://doi.org/10.1080/1474283032000139779

70. Stott, C. & Drury, J. Crowds, context and identity: Dynamic categorization processes in the 'poll tax riot'. *Hum. Relat.* 53, 247–273 (2000). https://doi.org/10.1177/a010563

71. Burns, D. *Poll Tax Rebellion.* (AK Press, 1992).

72. Fritsche, I., Jonas, E. & Kessler, T. Collective reactions to threat: Implications for intergroup conflict and for solving societal crises. *Soc. Issues Policy Rev.* 5, 101–136 (2011). https://doi.org/10.1111/j.1751-2409.2011.01027.x

73. Stollberg, J., Fritsche, I. & Bäcker, A. Striving for group agency: Threat to personal control increases the attractiveness of agentic groups. *Front. Psychol.* 6, (2015). https://doi.org/10.3389/fpsyg.2015.00649

74. Greenaway, K. H., Cruwys, T., Haslam, S. A. & Jetten, J. Social identities promote well-being because they satisfy global psychological needs. *Eur. J. Soc. Psychol.* 46, 294–307 (2016). https://doi.org/10.1002/ejsp.2169

75. Vansteenkiste, M., Ryan, R. M. & Soenens, B. Basic psychological need theory: Advancements, critical themes, and future directions. *Motiv. Emot.* 44, 1–31 (2020). https://doi.org/10.1007/s11031-019-09818-1

76. Fiske, S. T. *Social Beings: Core Motives in Social Psychology.* (Wiley, 2008).

77. Pittman, T. & Zeigler, K. Basic Human Needs. In *Social Psychology: Handbook of Basic Principles* (eds. Kruglanski, A., Higgins, E. T.) 473-489 (The Guilford Press, 2007).

78. Williams, K. D. Ostracism: A Temporal Need-Threat Model. In *Advances in Experimental Social Psychology* (ed. Zamma, M. P.) 275–314 (Elsevier Academic Press, 2009). https://doi.org/10.1016/S0065-2601(08)00406-1

79. Baumeister, R. F. & Leary, M. R. The need to belong: Desire for interpersonal attachments as a fundamental human motivation. *Psychol. Bull.* 117, 497–529 (1995). https://doi.org/10.1037/0033-2909.117.3.497

80. Cooke, A. N., Fielding, K. S. & Louis, W. R. Environmentally active people: The role of autonomy, relatedness, competence and self-determined motivation. *Environ. Educ. Res.* 22, 631–657 (2016). https://doi.org/10.1080/13504622.2015.1054262

81. Hamann, K. R. S., von Agris, A.-S. & Markus, L. Investigating the predictors of collective action intensity and health. https://osf.io/preprints/psyarxiv/qev28_v1 (2023).

82. Asah, S. T. & Blahna, D. J. Motivational functionalism and urban conservation stewardship: Implications for volunteer involvement. *Conserv. Lett.* 5, 470–477 (2012). https://doi.org/10.1111/j.1755-263X.2012.00263.x

83. McDougle, L. M., Greenspan, I. & Handy, F. Generation green: Understanding the motivations and mechanisms influencing young adults' environmental volunteering. *Int. J. Nonprofit Volunt. Sect. Mark.* 16, 325–341 (2011). https://doi.org/10.1002/nvsm.431

84. Ryan, R. L., Kaplan, R. & Grese, R. E. Predicting volunteer commitment in environmental stewardship programmes. *J. Environ. Plan. Manag.* 44, 629–648 (2001). https://doi.org/10.1080/09640560120079948

85. Reznickova, A. & Zepeda, L. Can self-determination theory explain the self-perpetuation of social innovations? A case study of slow food at the University of Wisconsin-Madison: Self-determination and social innovation. *J. Community Appl. Soc. Psychol.* 26, 3–17 (2016). https://doi.org/10.1002/casp.2229

86. Check-In Generator – Fragen für bessere Meetings & Workshops. https://www.checkin-generator.de/

87. Random Question Generator – Random questions. https://thestoryshack.com/tools/random-question-generator/.

88. denkwerk. Check in tscheck.in. https://www.tscheck.in

89. Reicher, S. & Haslam, S. A. Beyond help: A Social Psychology of Collective Solidarity and Social Cohesion. In *The Psychology of Prosocial Behavior: Group Processes, Intergroup Relations, and Helping* (eds. Stürmer, S. & Snyder, M.) 289–309 (Wiley-Blackwell, 2010). https://doi.org/10.1002/9781444307948.ch15

90. Herriger, N. *Empowerment in der Sozialen Arbeit.* (Kohlhammer, 2020).

91. Iyer, A. & Achia, T. Mobilized or marginalized? Understanding low-status groups' responses to social justice efforts led by high-status groups. *J. Pers. Soc. Psychol.* 120, 1287–1316 (2021). https://doi.org/10.1037/pspi0000325

92. Renger, D. & Simon, B. Social recognition as an equal: The role of equality-based respect in group life. *Eur. J. Soc. Psychol.* 41, 501–507 (2011). https://doi.org/10.1002/ejsp.814

93. Kragt, D. & Holtrop, D. Volunteering research in Australia: A narrative review. *Aust. J. Psychol.* 71, 342–360 (2019). https://doi.org/10.1111/ajpy.12251

94. Bergman, C. & Montgomery, N. *Joyful Militancy: Building Thriving Resistance in Toxic Times.* (AK Press, 2017).

95. Conyers, J. L., Jr & Smallwood, A. P. *Malcolm X: An Historical Reader.* (Carolina Academic Press, 2008).

96. TED. Sarah Corbett: Activism needs introverts – TED Talk. https://www.ted.com/talks/sarah_corbett_activism_needs_introverts (2016).

97. vanDellen, M. R., Campbell, W. K., Hoyle, R. H. & Bradfield, E. K. Compensating, resisting, and breaking: A meta-analytic examination of reactions to self-esteem

threat. *Personal. Soc. Psychol. Rev.* 15, 51–74 (2011). https://doi.org/10.1177/10888 68310372950

98. Nakashima, K., Isobe, C. & Ura, M. In-group representation and social value affect the use of in-group identification for maintaining and enhancing self-evaluation. *Asian J. Soc. Psychol.* 15, 49–59 (2012). https://doi.org/10.1111/j.1467-839X.2011.01361.x

99. Blaine, B. & Crocker, J. Self-Esteem and Self-Serving Biases in Reactions to Positive and Negative Events: An Integrative Review. In *Self-Esteem: The Puzzle of Low Self-Regard* (ed. Baumeister, R. F.) 55–85 (Springer US, 1993). https://doi.org/10.1007/978-1-4684-8956-9_4

100. Doosje, B., Spears, R. & Ellemers, N. Social identity as both cause and effect: The development of group identification in response to anticipated and actual changes in the intergroup status hierarchy. *Br. J. Soc. Psychol.* 41, 57–76 (2002). https://doi.org/10.1348/014466602165054

101. Abrams, D. & Hogg, M. A. Social motivation, self-esteem and social identity. *Soc. Identity Theory Constr. Crit. Adv.* 44–70 (1990). https://doi.org/10.2307/2076221

102. Sheldon, K. M., Wineland, A., Venhoeven, L. & Osin, E. Understanding the motivation of environmental activists: A comparison of self-determination theory and functional motives theory. *Ecopsychology* 8, 228–238 (2016). https://doi.org/10.1089/eco.2016.0017

103. Simon, B. & Stürmer, S. Respect for group members: Intragroup determinants of collective identification and group-serving behavior. *Pers. Soc. Psychol. Bull.* 29, 183–193 (2003). https://doi.org/10.1177/0146167202239043

104. Leach, C. W. *et al.* Group-level self-definition and self-investment: A hierarchical (multicomponent) model of in-group identification. *J. Pers. Soc. Psychol.* 95, 144–165 (2008). https://doi.org/10.1037/0022-3514.95.1.144

105. Lorenzi-Cioldi, F. Group Status. In *Oxford Research Encyclopedia of Communication* (2017). https://doi.org/10.1093/acrefore/9780190228613.013.421

106. Simpson, B., Willer, R. & Feinberg, M. Radical flanks of social movements can increase support for moderate factions. *PNAS Nexus* 1, pgac110 (2022). https://doi.org/10.1093/pnasnexus/pgac110

107. Amenta, E., Andrews, K. T. & Caren, N. The Political Institutions, Processes, and Outcomes Movements Seek to Influence. In *The Wiley Blackwell Companion to Social Movements* (eds. Snow, D. A., Soule, S. A., Kriesi, H. & McCammon, H. J.) 447–465 (John Wiley & Sons, Ltd, 2018). https://doi.org/10.1002/9781119168577.ch25

108. Giugni, M. & Grasso, M. T. Economic Outcomes of Social Movements. In *The Wiley Blackwell Companion to Social Movements* (eds. Snow, D. A., Soule, S. A., Kriesi, H. & McCammon, H. J.) 466–481 (John Wiley & Sons, Ltd, 2018). https://doi.org/10.1002/9781119168577.ch26

109. Van Dyke, N. & Taylor, V. The Cultural Outcomes of Social Movements. In *The Wiley Blackwell Companion to Social Movements* (eds. Snow, D. A., Soule, S. A., Kriesi, H. & McCammon, H. J.) 482–498 (John Wiley & Sons, Ltd, 2018). https://doi.org/10.1002/9781119168577.ch27

110. Almeida, P. *Social Movements: The Structure of Collective Mobilization.* (University of California Press, 2019). https://doi.org/10.2307/j.ctvd1c7d7

111. Dillard, M. K. Movement/Countermovement Dynamics. In *The Wiley-Blackwell Encyclopedia of Social and Political Movements* (eds. Della Porta, D., McAdam, D., Snow, D. A. & Dillard, M. K.) (John Wiley & Sons, Ltd, 2013). https://doi.org/10.1002/9780470674871.wbespm134

112. Corner, A. *et al.* How do young people engage with climate change? The role of knowledge, values, message framing, and trusted communicators. *WIREs Clim. Change* 6, 523–534 (2015). https://doi.org/10.1002/wcc.353

113. Goldberg, M. H., Gustafson, A., Rosenthal, S. A. & Leiserowitz, A. Shifting Republican views on climate change through targeted advertising. *Nat. Clim. Change* 11, 573–577 (2021). https://doi.org/10.1038/s41558-021-01070-1

114. Proulx, T. & Inzlicht, M. The five "A"s of meaning maintenance: Finding meaning in the theories of sense-making. *Psychol. Inq.* 23, 317–335 (2012). https://doi.org/10.1080/1047840X.2012.702372

115. Choi, E. U. & Hogg, M. A. Self-uncertainty and group identification: A meta-analysis. *Group Process. Intergroup Relat.* 23, 483–501 (2020). https://doi.org/10.1177/1368430219846990

116. Brewer, M. B., Hong, Y. Y. & Li, Q. Dynamic entitativity. *Psychol. Group Percept.* 19, 19–29 (2004).

117. Lickel, B. *et al.* Varieties of groups and the perception of group entitativity. *J. Pers. Soc. Psychol.* 78, 223–246 (2000). https://doi.org/10.1037/0022-3514.78.2.223

118. Callahan, S. P. & Ledgerwood, A. On the psychological function of flags and logos: Group identity symbols increase perceived entitativity. *J. Pers. Soc. Psychol.* 110, 528–550 (2016). https://doi.org/10.1037/pspi0000047

119. Lund, A. Atomkraft? Nej tak. *ATOMKRAFT? NEJ TAK* https://www.atomkraftnej tak.dk/smilingsun/ (2021).

120. Winterman, D. The other smiley. *BBC Mag.* http://news.bbc.co.uk/2/hi/uk_news/magazine/4484642.stm (2005).

121. Rast, D. E., Gaffney, A. M., Hogg, M. A. & Crisp, R. J. Leadership under uncertainty: When leaders who are non-prototypical group members can gain support. *J. Exp. Soc. Psychol.* 48, 646–653 (2012). https://doi.org/10.1016/j.jesp.2011.12.013

122. Bongiorno, R., McGarty, C., Kurz, T., Haslam, S. A. & Sibley, C. G. Mobilizing cause supporters through group-based interaction: Group interaction and mobilization. *J. Appl. Soc. Psychol.* 46, 203–215 (2016). https://doi.org/10.1111/jasp.12337

123. Pandolfi, F. How to create an effective non-profit mission statement. *Harv. Bus. Rev.* https://hbr.org/2011/03/how-nonprofit-misuse-their-mis (2011).

124. Jans, L., Postmes, T. & Van der Zee, K. I. The induction of shared identity: The positive role of individual distinctiveness for groups. *Pers. Soc. Psychol. Bull.* 37, 1130–1141 (2011). https://doi.org/10.1177/0146167211407342

125. Ende Gelände. Action Consensus. *Ende Gelände.* https://www.ende-gelaende.org/en/action-consensus-2021/

126. McKnight, P. E. & Kashdan, T. B. Purpose in life as a system that creates and sustains health and well-being: An integrative, testable theory. *Rev. Gen. Psychol.* 13, 242–251 (2009). https://doi.org/10.1037/a0017152

127. Martela, F. & Steger, M. F. The three meanings of meaning in life: Distinguishing coherence, purpose, and significance. *J. Posit. Psychol.* 11, 531–545 (2016). https://doi.org/10.1080/17439760.2015.1137623

128. Ojala, M. How do children cope with global climate change? Coping strategies, engagement, and well-being. *J. Environ. Psychol.* 32, 225–233 (2012). https://doi.org/10.1016/j.jenvp.2012.02.004

129. George, L. S. & Park, C. L. Are meaning and purpose distinct? An examination of correlates and predictors. *J. Posit. Psychol.* 8, 365–375 (2013). https://doi.org/10.1080/17439760.2013.805801

130. Westoby, R., Clissold, R. & McNamara, K. E. Turning to nature to process the emotional toll of nature's destruction. *Sustainability* 14, 7948 (2022). https://doi.org/10.3390/su14137948

131. Iron & Earth. About Iron and Earth. https://www.youtube.com/watch?v=-YKUim Z1l-c (2017).

132. Deci, E. L. & Ryan, R. M. The 'what' and 'why' of goal pursuits: Human needs and the self-determination of behavior. *Psychol. Inq.* 11, 227–268 (2000). https://doi.org/10.1207/S15327965PLI1104_01

133. Fritsche, I. Agency through the we: Group-based control theory. *Curr. Dir. Psychol. Sci.* 31, 194–201 (2022). https://doi.org/10.1177/09637214211068838

134. Fritsche, I. & Masson, T. Collective climate action: When do people turn into collective environmental agents? *Curr. Opin. Psychol.* 42, 114–119 (2021). https://doi.org/10.1016/j.copsyc.2021.05.001

135. Wullenkord, M. C. Basic Psychological Needs and Autonomous Motivation: A Humanistic Perspective on Pro-Environmental Behaviour Change. In *Handbook on Pro-Environmental Behaviour Change* (eds. Gatersleben, B. & Murtagh, N.) 211–225 (Edward Elgar Publishing, 2023). https://doi.org/10.4337/9781800882133.00022

136. Quested, E., Thøgersen-Ntoumani, C., Uren, H., Hardcastle, S. J. & Ryan, R. M. Community gardening: Basic psychological needs as mechanisms to enhance individual and community well-being. *Ecopsychology* 10, 173–180 (2018). https://doi.org/10.1089/eco.2018.0002

3

MORAL BELIEFS AND EMOTIONS

DOI: 10.4324/9781003558439-4

DEFINING MORAL BELIEFS

"I vividly remember standing at the edge of the coal pit close to Hambach Forest for the first time. Looking down into this massive hole larger than the size of Paris, I just got so angry. How could it be that coal mining – a highly destructive, polluting, and even inefficient way of producing energy – was given priority over the preservation of a precious ecosystem? For me, there was no doubt that this was fundamentally wrong. It needed to stop."

Can you recall an experience that left you feeling similarly outraged? The account above details how the violation of an individual's moral beliefs spurred them towards taking action as a member of the German climate action group *Ende Gelände* (pictured in Image 3.1). In climate action contexts, moral beliefs are everywhere. They are reflected in climate strikers' loudly declared: "What do we want? Climate justice! When do we want it? Now!" Or when Vanessa Nakate, a climate activist from Uganda, writes that incorporating the element of justice requires recognizing the Global North's moral responsibility to intensify its efforts to reduce emissions.[1]

It takes courage to become aware of and express your moral beliefs and the anger that may accompany them. Indeed, moral beliefs are key contributors in

Image 3.1: Activist looking out over the Hambach open-cast mine during the Ende Gelände actions, Germany (2017).

Photo by Pay Numrich (CC BY-SA 2.0)

what makes us act, making it important to take a deeper look at them in the context of motivating collective climate action.

You're most likely already familiar with the concept of *personal* beliefs – after all, we all have them. When it comes to our decisions in the produce aisle, for example, some of us may believe that organic foods should take priority, while others of us may believe opting for locally grown foods is what matters most.

For some of our personal beliefs, we have a strong opinion about what is right or wrong with no gray area in between. These are *moral* beliefs, characterized by a so-called absolutist stance. To take another food-related example, a vegan's moral belief might be that eating animal products is fundamentally wrong, no matter the circumstances – there might be the feeling of a moral truth underlying the belief which is unreservedly non-negotiable and transcends contexts and time[2]. This is why moral beliefs can feel so self-evident that they're beyond explanation. For example, if an individual who holds true the idea that unnecessary harm is wrong is asked "why is it wrong to harm a person who has not harmed you?". For them, the answer might seem so obvious that all they can say in response is, "because it just is" – without truly answering the question.

Looking at the larger picture, our moral beliefs can be connected to different spheres and actors within the climate crisis. They can link to the climate crisis itself (Is climate protection a moral issue?), to you personally and others (Am I/ is this person acting morally correct?),[3] to groups and institutions (Is a group responsible for climate injustices?), or to collective climate action (Is a specific form of climate protest moral or immoral?).

Climate activists often report strong moral beliefs pertaining to these spheres and actors. In an interview study, one activist reacted to the state investing pension funds in fossil fuel companies as follows: "So the system is entirely wrong […], politicians, who have a responsibility for our common earth, so to speak, should act on the basis of the warning signals that are there".[4] While it's plain to see that this activist has a clear idea of what is right and what is wrong, it's important to bear in mind that we don't all share the same moral beliefs. Indeed, the very same situation can be perceived as fundamentally wrong by one person and as completely unproblematic by another. The politicians who made the decision to invest pension funds in fossil fuel companies might not have seen anything wrong in doing so, perhaps because, for them, the moral priority lies not in divesting from fossil fuels but in ensuring elderly people have the funds needed for a healthy retirement.

When it comes to undertaking the difficult task of understanding the moral beliefs of others, looking at people's core values can be a first step.

Core values

Our core values represent what is important to us in life,[5] and our moral beliefs can motivate us to promote or protect these values in concrete situations. Acting as guiding principles for how we choose to act, our core values can often take the form of rather abstract goals we are striving for, such as wanting to protect nature.

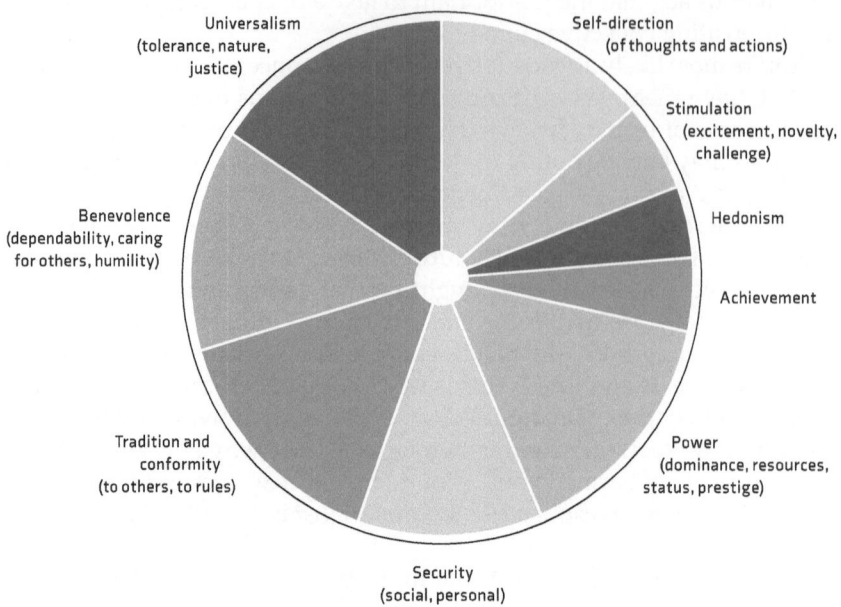

Figure 3.1: Value wheel, adapted from Schwartz and colleagues[7]

Researchers have long studied the concept of values. One such researcher, Shalom Schwartz from the Hebrew University of Jerusalem, developed the *Theory of Basic Values*, which has since become one of the most prominent frameworks for assessing values.[6] After analyzing data from 20 different countries, Schwartz and his research team came up with a set of 10 distinct core values that are important across cultures and contexts. They recognized that individuals within those contexts may have values that differ, of course. These ten values are represented as a value wheel in Figure 3.1.

How we prioritize these values in our lives is different for each of us. Some of these values and how they interconnect, as well as related goals, are explored in further detail in the following sections, along with examples of how they can be used in climate communication.

Values that go beyond our personal selves

Some of us may prioritize values that transcend our personal selves. For instance, we might be guided by values of universalism as the desire for a healthy nature as well as tolerance towards all people and justice for all (see Figure 3.1).[7] Many of those who are active in the climate movement or act in environmentally-friendly ways embrace these universalistic values.[8-12] As core values are often reflected in the mottos and mission statements of climate action groups, we can often learn about the goals of a group by looking at these elements. Take, for example, the group *Friends of the Earth*, who define themselves as being "a bold voice for justice and the planet".[13] This statement makes it clear that universalism is a prioritized value for this group.

Within social movements, you will likely encounter individuals who endorse values of benevolence. Such values are: caring for others, being dependable, and staying humble.[7]

Values that focus on autonomy and excitement

Some of us may prioritize values that connect to our desire for autonomy.[7] For instance, we might consider it essential that we have the ability to discuss freely with others and express our thoughts. One climate action group that reflects this value is *Transition Town*, who describe themselves as "a movement of communities coming together to reimagine and rebuild our world".[14] For this group, self-directed action as well as imagination is central to working towards the larger goal of global change.

We might also prioritize novelty, excitement, and the latest challenge.[7] One example of this is travelers wanting to show everyone that it is possible to travel the world by hitchhiking or using only solar energy.[15,16]

Values that concern self-enhancement and stability

Some of us may prioritize values in life that concern our own self-enhancement.[7] For instance, we might strive for achievement and success, power over others and over relevant resources, or a hedonistic lifestyle. Although sometimes frowned upon as not typically serving the cause, these values also exist within the climate movement; for example, the research psychologist who puts all her effort into her career success in order to prevail in a highly competitive system. Or the activist taking a gap year to follow her hedonistic passion for art.

Then there are people who value stability and maintaining the status quo; they place importance on cultural and religious tradition, conformity to others and to rules, and personal as well as social security. For instance, a member of *Churches for Future* may build their engagement on preserving Creation.

The value wheel

Of course, none of us focuses on just one single value. We all have a compilation of many values, and, importantly, we differ in how we prioritize these values in our lives. Some we endorse to a great degree and others to a lesser degree. Our core values can be placed on a wheel like the one shown in Figure 3.1. The closer one value is to another value on this wheel, the more likely an individual is to prioritize both values.[5] We can see that an individual who prioritizes nature (universalism) is also likely to prioritize forging their own path (autonomy) and less likely to prioritize dominating others (power).[5]

Reflecting on core values helps us to learn not only about ourselves but also about others. This can be particularly useful in attempting to understand why some people are attracted to climate action groups while others are repelled by them and their values. It may also help in understanding dynamics within climate action groups and the climate movement at large.

Our values, the values of others, and the values of society at large are not completely set in stone. Values can change over time. As one example, a large global survey on the COVID-19 pandemic's impact on values found an emerging shift away from materialistic values, which emphasize financial and physical security, to post-materialist values, which emphasize autonomy and freedom of choice.[17]

FROM MORAL BELIEFS AND EMOTIONS TO COLLECTIVE CLIMATE ACTION

Scholars call social identification and moral beliefs "the two chambers of the beating heart" of people involved in social and environmental justice movements.[18] Several studies in the broader field of collective action research have shown that our moral beliefs can motivate us to take action.[18] This is also true for collective climate action.[19-21] One study on climate protesters in Switzerland, for example, found that the protesters had stronger moral beliefs regarding the necessity of climate action than non-protesters did.[22]

Moral beliefs can have a motivational power in and of themselves, as they give us a sense of purpose.[23-25] In one study, members of the environmental movement in Norway reported that aligning their actions with their values also boosted their self-esteem.[23] But moral beliefs don't automatically translate into action. Our moral beliefs are not always front and center, and we can probably all recall situations in which we acted out of line with them.

> Moral beliefs often come into play when we perceive
> them as being violated.

Indeed, several studies on individuals who felt their moral beliefs had been violated by climate-related issues such as earthquakes induced by gas extractions, unequal access to clean water, the environmental situation in Iran, and US President Trump's withdrawal from the Paris Accords found that these violations were a motivating factor in their participation in collective action.[26-30]

For another example of how violated moral beliefs can motivate us into action, let's travel back to 2022, when the publication of a single article led to seismic outrage across the world. In the article, published by *Yard*, it was reported that in the first seven months of the year the average private jet usage among celebrities had emitted 482 times more than the average citizen's annual greenhouse gas emissions.[31] The report spread like wildfire, and readers reacted with incomprehension and outrage, expressing their feelings on social media.

One user pointed out that while the general public is advised to avoid driving and often faces criticism for commercial flying, wealthy individuals seem to have the freedom to do as they please. She further noted that one celebrity had even joked about using a private jet to fly her dog to her location because she missed it. To her, this behavior was unacceptable.[32]

It didn't take long before citizens started turning their anger into action. That year, climate groups in more than 13 countries launched protests calling for bans on private jets. And in 2024 a Europe-wide petition called for the same.[32]

Table 3.1: Examples of group-based inequalities

Type of inequality	Example for social group A	Example for social group B	Perceived injustice
geo-graphical inequality	Fijians (from the Global South), who are especially vulnerable to climate-induced sea-level rise[a]	those living in the Global North	the lifestyles of those living in the Global North contribute more to climate change, yet Fijians are more affected by it
gender inequality	women affected by natural disasters	men affected by natural disasters	structural inequalities make it harder for women than men to leave areas prone to climate disasters[a]
gener-ational inequality	younger generations, whose lives will be the most affected by climate change	older generations, who have contributed the most to climate change	Germany's constitutional court in 2021 ruled climate protection laws to be partly unconstitutional because they are threatening the fundamental freedoms of future generations[b]

a.   IPCC (Ed.) *Climate Change 2022 – Mitigation of Climate Change: Working Group III Contribution to the Sixth Assessment Report of the Intergovernmental Panel on Climate Change.* (Cambridge University Press, 2022). https://doi.org/10.1017/9781009157926

b.   BVerfG, 1 Senat. *Beschluss des Ersten Senats vom 24. März 2021 - 1 BvR 2656/18 -, Rn. 1-270.* (Bundesverfassungsgericht, 2021). https://www.bundesverfassungsgericht.de/SharedDocs/ Entscheidungen/DE/2021/03/rs20210324_1bvr265618.html

The moral belief violation chronicled in this example stems from a perceived injustice, namely that wealthy people are held to a different standard. When social groups such as those defined by race, gender, age, or sexual orientation are held to disparate standards or have disparate privileges, perceived injustice is a common outcome. Indeed, our collective history has countless examples of this as can be seen in Table 3.1.

Importantly, injustices and violations of moral beliefs are not perceived only through rational considerations and thoughts. Quite the contrary. Two meta-analyses have shown that emotions like anger, outrage, and resentment of injustice are prolific among those involved in collective action.[18,33]

Since feeling and thinking usually go hand in hand, we as the Author Team go along with the concept of *feeling-thinking processes* around injustice,[34] rather than attempting to disentangle thoughts from feelings. If a person prioritizes nature (universalism value) and holds related moral beliefs, they likely perceive a moral violation. This moral violation may lead to strong feelings and perceptions of injustice.[24,35,36] And this feeling of injustice might just turn them towards collective action. An example of what this process might look like in the context of climate injustice can be seen in Figure 3.2.

We also saw an example of this process taking shape in the quote from the first section in this chapter. That quote, taken from an anti-coal mining activist,

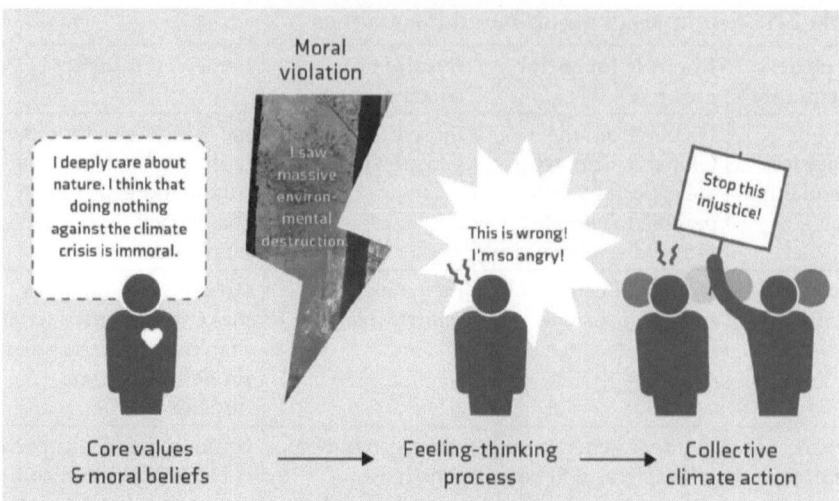

Figure 3.2: Pathway from moral beliefs and core values to action.

Photo of burning forest by Nikolay Kondev, 2019 (CC0)

demonstrates how someone can go from emotion to action. Based on the activist's desire to prioritize the "preservation of a precious ecosystem", we can glean that one of their moral beliefs is that nature must be protected. From the lines, "looking down into this massive hole [...], I just got so angry" and "this was fundamentally wrong", we know they perceived this to be a violation of their moral beliefs and reacted emotionally to this. And from the last line, "for me, there was no doubt that [...] it needed to stop", we know this violation spurred them into action.

While many emotions have the potential to motivate us to act, we will zoom in on two of them: guilt and anger. Let's take a deeper look at how the emotional reactions of guilt and anger can motivate people to take collective climate action.

Guilt

Personal guilt arises from feeling responsible for wrongdoings.[37] In the context of the climate crisis though, collective guilt, which arises from feeling like our social groups as a whole are responsible for wrongdoings, is particularly important.[38]

While research is somewhat limited, there are a few studies that found that people who felt collective guilt were more likely to want to join a neighborhood climate initiative[39] and that collective feelings of guilt were a source of social identification for members of *Fridays for Future*[40]. A guilty conscience may be uncomfortable, but if perceived on a collective level it may drive urgently needed action.

Anger

There is a greater amount of research on anger. For reasons of simplification, we will not differentiate between different forms of anger, such as (moral) anger, moral outrage, or indignation, and will use the terms interchangeably.[24,34] Anger is one of the six basic emotions.[41] It has a place in promoting adaptation – anger tells us something needs our attention and that we might need to act. In particular, anger can draw our attention to the fact that our moral beliefs are being violated, which can motivate us to act and stand up for our principles. Yet, anger often has quite a negative image as a hostile emotion.[42] Sometimes anger is even perceived as an immoral emotion,[36] which is somewhat ironic for a moral emotion.

Despite its negative image, anger can be an important motivation for collective climate action. Surveys of young people around the world found anger to be a common emotional reaction to the climate crisis,[43] and interviews with climate activists found anger to be a consistent driver for staying committed to the cause.[4,23,44] Anger was also strongly associated with environmental and climate-related collective action such as in studies on *Extinction Rebellion* actions, *Fridays for Future* protests, and environmental conflicts in Vietnam.[19–20,45–47] In one study on *Extinction Rebellion*, members who reported feeling anger over how people are treating the environment were more likely to identify strongly with the group and participate in collective action.[47] When compared to other emotions (sadness, hope, and guilt, among others), another study found that anger was most strongly related to getting people interested in joining a climate protest.[48]

While the mentioned studies found anger to be an important emotion for collective climate action, other research paints a more mixed picture. Studies on people involved in and sympathizing with *Hambach Forest* protests and *Fridays for Future* found that while anger was associated with (normative) collective climate action it was not as important as several other influencing factors.[49,50] Similar results were also found in studies on transition towns and neighborhood initiatives.[21,39,42] In yet another study, researchers managed to increase participants' anger by telling them that public anger about climate inaction was growing.[51] However, it did not lead to an increase in willingness to participate in collective action.

In an overview article, members of the Author Team argue that the mixed findings on the role of anger in collective action may be influenced by the specific type of action being studied.[52] Notably, anger appears to be more relevant in protest contexts than in other collective action contexts, such as neighborhood initiatives. Furthermore, the targets of this anger seem to affect its relevance. Interestingly, in studies assessing how angry people were at the government, anger was least relevant. In contrast, anger directed at people's qualities ("people don't care") and actions ("people don't do anything") seems to be particularly relevant.[48] Thus, while groups such as *Fridays for Future* rightfully address the actions of governments, it may be the actions of other people that more profoundly elicit outrage.

Box 3.1: Note – The gender inequality of anger

Many climate protesters identify as women, and many of these women are undoubtedly angry. Yet, research shows that women are less likely than men to express their anger about the climate crisis.[48,53,54] One reason for this imbalance could be the stereotypical view of anger as a male-owned emotion.[55] Accordingly, people perceive angry women differently from how they perceive angry men – when a man expresses anger, people tend to look for the external factors causing or contributing to it; when a woman expresses anger, it is more likely to be attributed to her personality.[56] In other words, anger experienced by a woman is seen as being rooted in her head and heart, whereas anger experienced by a man is seen as being rooted in the real world. What is more, women who express their anger are seen as having a lower status.[56]

In the climate movement, this unequal perception of anger may make it more difficult for women to spread messages of climate injustice, to express themselves freely, or even acknowledge what they're feeling.

With all that we've now learned about how moral beliefs and emotions can motivate people to act, let's take a look at how we can use these elements as tools for promoting and generating support for collective climate action.

HOW WE CAN USE MORAL BELIEFS AND EMOTIONS AS TOOLS FOR EFFECTING COLLECTIVE CLIMATE ACTION

As we've seen, the process of going from having moral beliefs to actively fighting to defend them involves multiple steps. This means there are numerous points at which we can attempt to draw on moral beliefs and emotions to promote collective climate action. The first part of this section focuses on just that, by providing strategies for how to create situations that elicit anger.

For the climate movement as a whole though, it's not only the moral beliefs and emotions of involved individuals that are worth focusing on. How collective climate actions are designed and implemented can make the difference between an action being seen as morally justified by bystanders and the public at large or being seen as a moral violation. And since moving the climate movement forward requires winning support from these people, the second part of this section offers strategies for using moral beliefs and emotions when talking to people outside the movement.

Focus 1: Creating anger-eliciting situations

Through experiencing feelings of anger, we can be spurred on to take collective action. For this reason, creating situations that tap into this anger can be a good strategy for motivating individuals to get involved in collective climate action. So, let's take a look at two strategies for how to do just that.

Before heading into these strategies though, you might want to check out Box 3.2, which will help you identify the things that provoke anger, a skill which may be useful to carry with you as you delve into the strategies for Focus 1.

 Box 3.2: Take action – Angry at the world

As you engage with collective climate action, you might encounter individuals harboring feelings of anger. On the other hand, you might encounter individuals whose feelings of guilt overshadow any anger they might also be feeling. Regardless of whether or not a person's anger sits on the surface or lies deep down, being able to harness that emotion can be a powerful skill for motivating someone to take action. And the first step in developing this skill is learning how to isolate what provokes anger.

Think about where you see expressions of anger within the climate movement. What are the targets for the anger being expressed by…

… *your friend who engages in climate action?*
… *Greta Thunberg?*
… *the group members of a climate action group of your choice?*
… *your climate-related role model?*
… *you?*

If you struggled to pinpoint the targets for the anger being expressed by these individuals, try discussing them with others. Or, if you're looking for a unique source of inspiration, listen to "Angry at the World" by *The Kyteman Orchestra*, a favorite song of one of our authors.

Focus 1 – Strategy 1: Drawing attention to injustice and who's responsible for it

The activity in Box 3.2 brings us to our first strategy. While it might be difficult to instill new moral beliefs in others, emotions can be elicited through various tactics, such as drawing attention to injustices and the people or institutions behind them. One way to do this is strategic communication through messages and campaigns.

In one experimental study on strategic communication, participants were given one of two texts to read. One text focused on Germany being a major contributor to global pollution, while the other focused instead on Germany being a responsible agent in the field of environmental protection.[37] As might be easy to imagine, participants who were assigned the first text reported stronger feelings of guilt and anger than those who were assigned the second.

In a similar study, a group of researchers had participants watch one of two videos.[49] One video presented viewers with scientific facts about coal mining practices. The other comprised clips from real footage collected by an anti-coal group. This video showed the unfair practices of coal mining companies as well as the politicians who enable them. Participants who were shown the second video reported stronger feelings of anger than those who were shown the first. They also reported stronger agreement with the idea that those responsible for the unfair practices should be held accountable. What's more, they even reported stronger feelings of being moved, which, in turn, related to support for the forest protection campaign in this coal-mining region.

Sharing videos, like the one mentioned in the study, across various social media platforms is a frequently used tactic by climate action groups looking to draw attention to issues and mobilize support. When thinking about how best to do this, a communication strategy that involves highlighting a single, specific injustice and, if possible, focusing on who might be behind that injustice might be the most effective way of eliciting anger and, consequently, motivating people to act. This concept is nicely illustrated by Greta Thunberg:

> We need to hold the people in power accountable for what they have been doing to us and future generations and other living species on earth. And we need to get angry [...] and then we need to transform that anger into action and [...] stand together united and just never give up.[57]

An organization that lays the foundation for anger by dedicating itself to exposing injustices is *Urgewald*. This organization publishes the Global Coal Exit List, which is composed of the names of companies involved in the coal industry worldwide – from coal power producers to coal transporters and traders.[58] *Urgewald*'s goal with this list is to provide a single, clear source for finding out which companies are most responsible for CO_2 emissions and thus any related climate injustices. The list is also a resource for environmentally-conscious investing.

Showing the consequences of climate change is another way of drawing attention to injustices. For instance, the Maldives Islands are at risk of disappearing by the end of this century due to rising sea levels.[59] As shown in Image 3.2, the Maldives Islands' cabinet held the world's first underwater cabinet meeting in 2009. This meeting was a powerful demonstration of how they are disproportionally affected by climate change.

 Box 3.3: The bottom line

Communication strategies that highlight a specific injustice and who's behind that injustice can be effective at evoking emotions like anger and guilt. In turn, these emotions could motivate an individual to fight these injustices on a collective level.

Image 3.2: Fisheries and Agriculture Minister Ibrahim Didi signs the decree of the underwater cabinet meeting off Girifushi Island, Maldives (2009).

Mohamed Seeneen (CC BY-NC-ND 2.0)

Focus 1 – Strategy 2: Communicating others' outrage

Showing how others are already outraged can be a powerful lever to draw attention to climate injustices. Two approaches seem useful here: highlighting emotions as norms in social groups and considering how different levels of being affected by climate injustice can make a difference to the expression of anger. Let's take a look at them.

The role of outrage norms

Social norms represent our perception of what others do or don't do, what they approve or disapprove of. Whether or not we're always aware of it, social norms influence our actions (see Chapter 2 for more on social norms). In a similar way, our assumptions of how others feel can have an impact on us. When we perceive that many others are angry, we perceive strong outrage norms. In one experimental study on communicating the feelings of others, researchers had participants read a text about the prevalence of waterborne diseases in developing nations as well as the *Water for Life* movement, which fights to make clean drinking water a universal right.[60] After reading these texts, participants were asked to come up with strategies for how to motivate people to support the movement in small groups. One group was told that feelings of outrage motivate people to support the movement, thus implying outrage was a prevalent emotion and norm within this group. The study found that the participants who received this additional information about an outrage norm priming their group-based discussion experienced greater feelings of outrage than those who did not receive it. What's more, they reported feeling more motivated to act themselves. What this study tells us is that communicating the idea that outrage is the norm within a given space or situation might be useful when trying to elicit feelings of anger.[61]

Sharing survey data as a way of providing a picture of just how many people are outraged about the environment is also an effective way of provoking anger and garnering support. For example, the German Federal Ministry for the Environment, Nature Conservation, Nuclear Safety and Consumer Protection regularly runs representative surveys in Germany. In 2022, a survey found that 98% of respondents reported feelings of outrage over human-made environmental problems,[62] which is more than you might have expected.

Hearing about the feelings of others lets us know that we are not alone. Indeed, we might come to find that anger is the one emotion that seems to be shared by many others. Learning about others' feelings may also give us the opportunity to confront and build stronger connections with our own emotions, values, and moral beliefs.

At meet-ups, climate action groups might benefit from discussing how members' feelings of anger interact with their moral beliefs and values. These discussions might give members who have yet to process their anger a way to ground the emotion in their values or moral beliefs, thus making the anger seem effective or positive. And members who feel stressed or overwhelmed by their anger can engage in discussions about appropriate ways of handling and channeling it.

In one study that has yet to be published, the Author Team surveyed people involved in the movement for socio-ecological transformation. Surprisingly, this study found that when various factors were simultaneously considered, respondents who held stronger moral beliefs regarding socio-ecological topics committed *fewer* hours than others to their climate action groups.[63] One way to explain this finding might be that individuals with particularly strong moral beliefs are more predisposed to conflict, which in the case of group participation might lead to a negative impact on their levels of commitment. To help mitigate this, it might be constructive for groups to actively establish meeting structures and group dynamics that enable members with various moral beliefs and varying strengths of feelings to participate without feeling threatened or being perceived as threatening.

Global North/Global South divide – Communicating outrage from different perspectives

Certain groups of people may be more willing than others to voice their anger (such as mentioned in Box 3.1). In one interview study on global climate activism, activists from the Global North reported feeling less inclined to give anger a prominent position in their mobilizing activities than activists from the Global South.[4] Those from the Global North reported seeing value in treating anger with caution – they saw focusing on anger and accusations as fueling resignation and frustration instead of motivating action. Activists from the Global South on the other hand reported seeing value in utilizing anger – they saw focusing on the

Global North as the main driver of climate change as a way of motivating action. Allowing anger may therefore be easier for those affected by climate injustices – whether personally or as part of a specific group defined by region, income, generation, or gender.

Staying with the Global North/South divide, one real world example of holding an external group accountable comes from the climate action campaign *Debt for Climate*. The goal of this campaign is the cancellation of financial debts countries in the Global South have with countries in the Global North, the purpose of which would be to enable those indebted countries to pursue a self-determined socio-ecological transformation.[64] Many of these countries are forced into exacerbating the climate crisis because the only way of acquiring the vast amount of funds needed to pay off their debts is fossil fuel extraction. Initiated and led by people from the Global South, *Debt for Climate* brings the voices of those who are most affected by the climate crisis to center stage and provides a platform for highlighting climate injustice. The campaign also employs the strategy of establishing outrage as a social norm. Their promotional video features outraged protest speakers as well as further group members clearly stating that, "we demand the [International Monetary Fund] and World Bank to cancel all Global South debts and turn these funds into climate action!"[65] If such a goal sounds unattainable to you, it is worth noting it has historic precedence in the Global North. In fact, one prominent historical example is the cancellation of Germany's debt after World War II, which allowed for its quick economic recovery.

 Box 3.4: The bottom line

Communicating that outrage is widespread and helpful for collective action can help more people access their own anger and feel motivated to act. People who are personally and collectively most affected by the climate crisis may be particularly open to expressing their anger.

Focus 2: Using moral beliefs and anger to build external support

Thus far, we've seen a lot of ideas about how to motivate individuals and social groups to get involved in collective climate action as well as ideas for how to maintain motivation among members of climate action groups. And yet, even with these strategies, there will always be those who remain simply unwilling or unable to dedicate their time, energy, and resources to collective climate action, regardless of any amount of targeted effort to the contrary. What's more, some research even shows that there are those who are most angry not about the climate crisis but about the policies trying to mitigate its effects.[48] Of course, there are those who remain unconvinced the climate crisis is even really an issue at all.

And then there is another group whose influence we shouldn't underestimate: the supporters; those individuals who see potential in showing support to the movement but are happy to maintain an existence outside it. Indeed, building public momentum is considered an essential element of true societal transformation.[66-68] Scholars argue that increased public support can correspond to increased financial support and increased political influence.[69] There are numerous examples of social movements that were able to use public support to achieve substantial political change despite having relatively low direct participation rates.[70] What all this tells us is that,

> public support, however passive, can make the
> difference between a successful movement and an
> unsuccessful one.

And this raises the question, when it comes to organizing a successful movement, how can we build support among those who might never actively engage in collective action themselves?

Focus 2 – Strategy 1: Narrowing the moral-empathy gap with direct contact

While it may feel perfectly natural to encounter differences in opinions when discussing where to spend the holidays or what pizza toppings to get, encountering differences in moral beliefs often feels noticeably unnatural and even upsetting. This stems from the idea that our moral beliefs have a way of becoming somewhat factualized for us – we feel that our beliefs are "right" and that everyone else should adhere to them as well[2]. One result of this is that we can end up viewing the moral beliefs of others as "just wrong" when they are incompatible with our own beliefs, which makes it hard to not only understand but also empathize with the people who have those moral beliefs. This struggle is known as the moral-empathy gap.[71]

An example of a moral-empathy gap was observed in one study of participants in a non-normative climate action.[71] The activists surveyed thought that their actions would increase rather than decrease public support, but the reality was the opposite. People who were not involved viewed the actions as immoral and did not support the activists – a possibility that may have puzzled the activists, whose moral beliefs likely justified and even necessitated the non-normative action.

This case highlights how difficult it can be for members within a movement to understand the perspectives of those outside it. And since we've already learned how integral public support can be in the success of a movement, let's take a look at ways climate action groups can narrow the moral-empathy gap and become more aware of how others perceive climate actions.

Engaging with outsiders in everyday life

It might be helpful to engage with views other than your own in everyday life. With specific regard to climate action, you might encounter disparate perceptions of certain protests or marches in the daily paper, on the news, or in social media. Where possible, engaging with the people who hold these disparate views might have the dual success of improving both their understanding of your perspective and your understanding of theirs. Bringing this skill to the climate movement might mean being better able to share information on collective action with outsiders, which has the potential to increase public support and therefore the success of a movement.

The campaign *Deutschland spricht!*[72] [Germany speaks] is one example of people with differing views coming together, creating an opportunity to bridge the moral-empathy gap. Participants in the campaign were asked to fill out a survey on their currently-held political opinions, after which they were matched with a fellow participant with dissimilar views. After meeting with their political opposite, the majority of participants reported feeling satisfied with their interaction.[72] While nearly three-quarters of participants reported their perspectives had been reinforced by the meeting, around half of participants also reported having been convinced to change perspectives on one or more points.

Engaging with outsiders during collective actions

When it comes to garnering public support, it can also be a worthwhile endeavor to engage with outsiders while participating in collective actions. During a collective climate action, clashes of perspective are likely to occur between those involved and those bearing witness (bystanders) or directly affected by the action. Let's take a look at one example of how this might play out in the non-normative action of blocking a road in Table 3.2.

As we can see, the two groups of people in this example are not aligned in their perceptions of the road blockage. Whether or not these two groups share other values or moral beliefs we do not know, because any potential for finding common ground is lost in the intense reaction of those affected by the collective action.

Sticking with this example, there are a few tactics that people engaged in blocking a road could employ as a way of allaying public opposition. They could establish an outreach team to get involved in active dialogue with commuters. This would allow commuters to both voice their frustration with the

Table 3.2: Example of divergent moral perspectives

Groups	Moral perspective	Resulting emotion
People involved in collective action	"Desperate times call for desperate measures. The climate crisis is so grave, and no one seems to care, but if I block this road, they'll have to care. It's the moral thing to do."	Outrage at public inaction
People affected by collective action	"Actions like these only hurt the little people. These activists are keeping me from getting to my job – they don't care about regular people like me at all. This is morally wrong."	Outrage at the people involved in the action

Image 3.3: Extinction Rebellion leaflet apologizing for the inconvenience that upcoming action will cause and explaining the moral justification for it, UK.
Photo by Wandelwerk e.V.

inconvenience of having their commute disrupted and learn about the significance of the road blockage from the perspective of those involved.

The people involved could also preemptively share outreach material with those who are going to be most affected by the action. This outreach material might serve to share key details and information about the issue and how not enough is currently being done, thus somewhat justifying why such intense action is now going to be taken. An example of this kind of outreach can be seen in Image 3.3. These two tactics have the potential to foster empathy among people involved in collective climate action, bystanders, and those affected by the action – a necessary step towards bridging the moral-empathy gap.

Of course, even with these tactics, it's unrealistic to expect to change the mind of everyone, no matter how much we might wish that to be the case. Regardless of how much effort is put into outreach, climate action groups need to accept the possibility of enduring opposition. Indeed, there are plenty of situations in which walking away from a difficult conversation may be a more effective approach than continuing to try to change a person's mind – in these cases, the people involved in climate action can see success in having maintained their own mental health or deescalated a situation.

The downside to engaging positively with outsiders

When it comes to focusing our efforts on engaging positively with outsiders, whether in everyday life or during collective actions, it is worth acknowledging the potential for undesired consequences. Research from other social contexts has found that positive interactions between the members of a disadvantaged group and the members of an advantaged group can have the result of increasing the former's gratitude for the latter.[73] This in turn can lead to a decrease in the disadvantaged group members' motivation to engage in protests, as the fuel for their initial outrage has been somewhat reduced.

Extrapolating this research, we may also see the possibility of similar downsides resulting from positive interactions between people experiencing a climate injustice and the individuals or groups contributing to that injustice. For example, if more and more individuals from the Global South started having positive interactions with individuals from the Global North, which of course could have a lot of benefits, it might also cause the former to be less willing to focus their outrage on the latter. And, as we have already learned in this chapter, without a target for their outrage, individuals might lose their motivation to engage in collective action, which could have negative effects for the movement as a whole.

It's worth noting that a reduction in anger and subsequent reduction in willingness to engage in protest might open up doors to other kinds of engagement within a movement. This could include forms of participation that don't rely heavily on anger, such as neighborhood involvement. Overall, though, since positive interactions with bystanders, those responsible and the general public have the potential to both create empathy and reduce anger, it seems important to weigh these upsides and downsides when deciding whether or not to concentrate effort, and how much effort, on engaging with individuals outside the climate movement.

 Box 3.5: The bottom line

Members of climate action groups often have strong moral beliefs. Thus, it might be hard for them to understand and empathize with the perspectives of people outside the climate movement or of people who don't share the same moral beliefs and values. This difficulty is called the moral-empathy gap. Group members can work towards bridging this gap by concentrating efforts on engaging positively with outsiders in everyday life or with bystanders and those affected by an action as a way of fostering mutual empathy, understanding, and even support.

Focus 2 – Strategy 2: Challenging, not threatening, others' self-image

Since our moral beliefs contribute to laying the foundation of who we are and how we act, it is only natural that they come up in everyday life. As we engage with social media, colleagues, friends, and family, the moral beliefs we feel most strongly about are particularly likely to come up. However, natural though this may be, conversations that involve moral beliefs have a way of inciting frustration and defensiveness when people disagree. For instance, a person eating a vegan diet based on her moral beliefs may experience defensive reactions to her veganism. Even if she merely selects the vegan option in the cafeteria, her friend might reply with something like, "I would have picked the vegan food if it had looked more delicious today", or, "don't judge me for picking the beef burger".

The reasons behind reactive statements like these may not always be clear. One reason for defensiveness might be the need to alleviate guilt. In the described example, guilt could come from the possibility that the friend not choosing the vegan option actually does agree that consuming animal products is immoral, and in being confronted by their friend doing the moral thing, they felt the need to justify their choices, primarily to themselves. When we violate our own moral beliefs, we often experience feelings of guilt or shame, and instead of confronting these emotions, sometimes we react with defensiveness to whatever or whoever we see as the reason our moral violation was brought to our attention in the first place. In other words, our moral self-image becomes threatened.[74]

Researchers who have studied the concept of the moral self-image and what happens when it's threatened have developed a concept they call the *moral rebel*.[75] A *moral rebel* is someone who rebels against compromising their morals. In an experimental study on how *moral rebels* are perceived, researchers asked participants to individually do a task that had racist elements. Unknown to the participants, one among them doing the same task was actually a research assistant playing either a *moral rebel* or an obedient other. The *moral rebel* refused to carry on with the task on the grounds that the task was violating their moral values. Interestingly, how the *moral rebel* was evaluated by other participants depended on whether or not they carried out the task themselves. While uninvolved observers judged *moral rebels* positively for standing up for their principles, study participants who had previously carried out the problematic task themselves and only learned afterwards that the *moral rebel* refused to do the same disliked the rebel.[75] A possible explanation for this finding could be that involved participants felt threatened in their moral self-image because they did not refuse to carry out the task themselves.

So, what can we do when we see someone behaving in a way that is morally questionable? There are numerous approaches to handling this, ranging from doing nothing to employing research-backed strategies for helping a person regain self-efficacy and be motivated to change,[76] so let's take a look at just a few of them.

Challenging the action, not the character

A research-backed strategy posits that challenging a person's actions is more effective in motivating change than challenging their character is.[76] Let's take wanting someone to cut back on flying as an example goal. One way this strategy might play out is instead of making sweeping statements like, "anyone who takes frequent flights is a bad person", opt for statements like, "taking frequent flights has negative impacts on the environment". In the former, the focus is on the type of person who would engage in the morally-questionable behavior whereas in the latter the focus is on the behavior itself.

Sometimes, it may even be better not to focus on this particular person's actions but on the actions of others. Criticizing the behavior of others can be an effective way of helping a person reflect on their own behavior. When we draw attention to the immoral behavior of others outside of a given conversation, those within that conversation are given an opportunity to critically reflect on the issue at hand without feeling threatened in their own self-image.

For example, you notice how your mom often speaks negatively about climate activists, and you would like her to reflect on her opinion. Instead of confronting her directly, you tell her about a conversation you overheard the other day in a café where two people were making disparaging remarks about climate activists. They were congratulating themselves for sticking "Fuck you, Greta" stickers to their SUVs. You explain to your mom how such conversations just make you really sad. Disagreeing on certain forms of climate action is one thing, but overall, climate activists are striving for a better future for everyone. And they are investing considerable time and energy into this important work.

Reminding people that change doesn't have to happen quickly

Another strategy involves reminding people that change doesn't have to happen all at once.[76] Effective behavior change often involves a series of smaller goals and occurs over a period of time – it doesn't usually comprise one single ultimate goal or happen overnight. For example, for someone with the moral belief that animal rights are important and who has the goal of switching to a plant-based diet, it might be effective to help them focus on how they can align their moral belief with small, attainable goals on their way to achieving their ultimate goal of belief-behavior alignment. Diving straight into full belief-behavior alignment might leave this person more vulnerable to feelings of failure, guilt, and defensiveness.

Importantly, research shows that focusing on smaller goals works best if people don't use achieving these smaller goals as an excuse for regressing to immoral behavior – sticking with the example of a plant-based diet goal, an adverse excuse might look like, "I ate plant-based all week, now I can have some goat cheese". When reminding people that change can happen over time, it's therefore important to highlight overarching goals and values.[77]

It is useful to emphasize that skills are developed over time on a person's journey to promote climate justice.[76] If the goal of a particular conversation with a fellow urbanite is to convince the other person to align their moral beliefs with mitigating the climate crisis and to get rid of their preferred method of transportation (their gas-guzzling car) to help with this cause, it might be effective to share the story of someone who weaned themselves off a car over time by engaging more and more with public transportation or cycling before ultimately realizing they didn't actually need to use their car at all.

Table 3.3: *The four steps of non-violent communication*[78]

Step	Description
1) Observation	- describing specific actions or behaviors - describing situations without judgement or evaluation
2) Feelings	- identifying and expressing the feelings that occur in response to observed actions or behaviors: feeling content, grateful, free, happy, fulfilled, moved, enthusiastic, jolly, interested, peaceful, joyful, relaxed, cheerful, refreshed, anxious, depressed, numb, lonely, melancholic, angry, helpless, sad, bitter, spiritless, nervous, frustrated, desperate, worried, or gloomy - moving beyond "good/bad" and "right/wrong" descriptions
3) Needs	- recognizing the numerous universal human needs that can underlie feelings: authenticity, individuality, security, clarity, structure, self-esteem, appreciation, belonging, honesty, solidarity, mindfulness, attention, relaxation, harmony, inspiration, variety, happiness, celebration, learning, contribution, efficacy, growth, success, creativity, or meaning - avoiding focusing on or judging a person's strategies for satisfying these needs
4) Request	- making clear, positive, and actionable requests to improve your quality of life - accepting that the answer may be "no"

Engaging in non-violent communication

The aim of *non-violent communication* is to encourage empathic listening and open dialogue.[78] As such, it is another effective tool for narrowing down the moral-empathy gap. In *non-violent communication*, conflict resolution is fostered by shifting the focus away from judgement and blame.

A core assumption of this strategy is that all humans share similar needs, through which we can find mutual understanding.[78] Engaging in *non-violent communication* is a useful method for promoting empathy, compassionate relationships, and collaboration in group settings, by which the self-esteem of those involved is strengthened.[79-81] This type of communication focuses on four key steps, shown in Table 3.3.

To gain a better understanding of *non-violent communication* in action, let's take a look at a narrative developed by a *Wandelwerk* member:

At a family gathering one night, Uncle Howard, who knows that his niece, Mary, is involved in climate action groups, declares, 'I can't stand the CO_2 taxes'. Being familiar with the framework of *non-violent communication*, Mary listens empathetically and responds to Uncle Howard by asking, 'When you say you can't stand the CO_2 taxes (observation), do you mean they make you feel angry (feeling)? Or frustrated (feeling)? Is this about justice for you (need)? Are you concerned that paying these taxes will keep you from having enough retirement funds (need)? Or are you worried that things are changing too quickly for you to feel comfortable (need)?' Of course, Mary doesn't ask all these questions at once as she knows that might lead to Uncle Howard feeling attacked. But asking a few of them might signal to her uncle that she's trying to empathize with him. So, Uncle Howard can start to feel more secure and relaxed.

Having made sure the dialogue feels secure and relaxed, Mary now sees the possibility of sharing her own ideas on the topic. She says to her uncle, 'When you say that the CO_2 tax annoys you (observation), I myself start to feel angry (feeling). I see so much suffering in the news right now caused by the climate crisis (observation). I want to be able to trust that we are acting responsibly towards all people on this earth (need), and I would like all people to be able to live as freely as I do (need). I would ask you to try to understand where I'm coming from (request)'. Uncle Howard might be a bit taken aback at first, but at least he now has the chance to see what his niece really thinks. It is not unlikely that Mary will have to show her uncle empathy again as she proceeds from here.

The components of *non-violent communication*, as seen in this narrative, and the components of other concepts described in this book both overlap and disagree. While *non-violent communication* emphasizes and utilizes the diversity of feelings and needs as a method, academic psychologists often prefer to look for the most central feelings and the most basic needs under certain circumstances. What is also notable is that *non-violent communication* discards a right or wrong logic, something that moral beliefs are founded on. As well, *non-violent communication* seeks to put the needs that everyone shares at the center, while values are in turn pushed to the background, since they're more related to individual and collective strategies for satisfying people's needs.

All that being said, you may be wondering how effective *non-violent communication* is as a tool for motivating collective climate action or convincing people of the climate emergency. But it's important to keep in mind that motivating and convincing should not be the goals when using this kind of communication. The aim with *non-violent communication* is to build positive relationships with people, not to persuade them.

Nevertheless, using *non-violent communication* may help us bridge the moral-empathy gap, making it possible to understand each other's feelings and needs without needing to agree on them. Moreover, we can use the tools of *non-violent communication* to practice self-empathy in stressful situations, by reflecting on our own observations, feelings, and needs. For example, a member of the climate movement might feel uncomfortable in a conversation with a climate denier. Taking a moment to reflect on their discomfort and realize they're feeling frustrated and that their need for climate justice for all is not met may help them relax in an otherwise stressful situation.

Box 3.6: Take action – The moral-empathy gap from two perspectives

At this stage, you may want to think about situations in which you have experienced the moral-empathy gap yourself.

First, ask yourself:

• *Can you recall a situation in which you noticed another person's immoral behavior?*
• *How did you react? Did you point it out, and if so, how?*

Then, ask yourself:

• *Can you recall a situation in which you acted not in accordance with your moral beliefs, and you were confronted with that incongruity by someone else?*
• *How did that make you feel?*
• *How would you have wanted the other person to react?*

Reflecting on the moral responses you wish to receive may help you align with your moral beliefs about how people should communicate with each other.

Considering normative and non-normative protest

It is worth mentioning that, given the glaring injustices of the climate crisis, challenging people's morally-misaligned behavior through normative or non-normative protest is also a viable option. Collective actions can potentially inspire others to do the same, thereby contributing to the spread of protest actions as part of a larger societal transition.[76]

Of course, it is always up to you to decide if you prefer to actively protest or focus your efforts on social harmony and mitigating moral conflicts. And indeed, these decisions may vary under different circumstances and at different times. If you are considering getting involved in protest action though, the following sections provide information on the types of public impact achieved by various types of collective action.

 Box 3.7: The bottom line

Bringing up the immorality of climate issues can cause defensive reactions in people who feel that their self-image as a moral person is threatened. To counteract this effect, communication strategies could focus on behaviors instead of character, the immoral behaviors of others, and step-by-step adaptations instead of radical changes. Additionally, *non-violent communication* can be a useful tool for building and maintaining positive relationships even in disagreements.

Focus 2 – Strategy 3: Balancing public support and media attention: The activist's dilemma

In the fall of 2022, the streets and museums of Berlin became the backdrop for the protest actions of a small group of determined climate activists called *Letzte Generation* [Last Generation]. Berlin residents were in uproar over the group's actions, which ranged from glueing themselves to the streets to throwing tomato soup and mashed potatoes against glass-protected paintings in museums. *Letzte Generation*'s actions achieved two things: mass awareness and public opposition.

Within days of their first action, it seemed everyone was talking about *Letzte Generation*. Everyone. In taxis, on trains, at restaurants – people everywhere were expressing their opinions about these climate activists and what they'd done. While the public discourse seemed focused primarily on their actions and less on the actual climate issues, German polling data did show a peak in the perceived urgency of the climate crisis around the same time media coverage of *Letzte Generation*'s actions reached its peak[82]. What's noteworthy here is that the German news cycle was at the time dominated by the war in Ukraine and natural gas shortages, and yet despite all that, *Letzte Generation* was able to make headlines and bring attention back to the climate crisis.

But that attention wasn't necessarily favorable. Indeed, the public's reaction towards the protests was actually quite negative. The mayor of Potsdam, a city just outside of Berlin, described *Letzte Generation*'s act of throwing food at a Monet painting as "cultural barbarism".[83] Others called the group's actions unacceptable, high-risk, ruthless, and illegitimate.[84] Based on these assessments, we can surmise that many people outside the group perceived the actions as violations of their own moral beliefs. These negative perceptions of the group's actions coupled with the awareness these actions generated highlight a regularly occurring phenomenon within the collective action space.

> The phenomenon that a collective action can have both positive and negative consequences for public perception is referred to as the activist's dilemma.[71]

Perhaps the most relevant characteristic in how a collective climate action is perceived is whether it is normative, such as a peaceful demonstration, or non-normative, such as civil disobedience. A recent series of non-climate related studies explored the role of morality in the context of non-normative extreme protest actions and its effect on movement support.[71] In these studies, extreme non-normative protest actions were defined as those which observers perceived as harmful to others, disruptive, or both. Examples of this kind of action included freeing animals, protest chants encouraging violence against police officers, and blocking the entrance of an abortion clinic. These studies found that the extreme protest actions were rated as more immoral compared to the more moderate protest actions. What's more, the protest actions rated as more immoral led to decreased emotional connection, reduced social identification with the movement, and less support for the movement among study participants. In line with this, studies generally find that people outside a given movement show less support for non-normative protests than for normative protests.[71,85-87]

Overall, what these studies show us is that certain protest actions might undermine public support for the climate movement. However, most groups don't engage in extreme protest actions without reason, and climate action groups are no different – the climate crisis is worsening at an alarmingly fast rate. And more non-normative protest actions typically attract more media and public attention.[88-90]

Thus we come to the crux of the activist's dilemma: while non-normative collective actions can attract attention or put pressure on institutions, they can at the same time undermine public support for the social movement if the protest action is regarded as immoral.[71] It is therefore crucial before engaging in any kind of protest action to first determine what the goal of the action is. If it's to attract attention, non-normative collective action has proven to be more effective. If it's to gain support, normative collective action seems more promising. Of course, climate action groups often attempt to pursue both of these goals with a single action, which is why the following strategies focus on understanding the impacts of an action as well as how to balance those impacts.

 Box 3.8: The bottom line

The activist's dilemma stems from the fact that non-normative collective actions might simultaneously have positive and negative effects for the climate movement. Specifically, they can lead to increased media attention but carry the risk of losing public support. Therefore, it's vital to determine the goal of a climate action before enacting it.

Focus 2 – Strategy 4: Considering the impacts of a radical flank

Within most social movements, there is at least one subgroup that is more extreme than its other more moderate counterparts. This faction is often referred to as the radical flank, and it has a way of attracting significant attention. One of the most prominent examples of a radical flank can be taken from the pages of US history – the civil rights movement's *Black Panther Party*.

Spanning just over a decade in the mid-20th century, the civil rights movement campaigned for the abolition of racial segregation, discrimination, and disenfranchisement nationwide. On one side of the movement, the civil rights mainstream, led by prominent figures like Martin Luther King Jr. and Rosa Parks, fought for racial equality through peaceful protest, civil disobedience, and legal challenges. On the other side of the movement, radical groups with key figures like Malcolm X and the members of the *Black Panther Party* fought to achieve the same aim but through direct action, self-defense, and radical changes to the system.[91] This side of the movement was founded on the belief that non-violence and peaceful protests were not enough to bring about real, lasting change.[91] While both groups were dedicated to the goals of racial equality and justice, a clear line existed separating them into a moderate faction and a radical flank.

Some of you may now be asking yourselves whether or not your climate action group is a radical flank. The radicality of a faction is always defined in relation to its more moderate overarching group, and flanks can appear in various shapes and sizes. A radical flank typically uses more extreme forms of action (not necessarily violent or illegal), and also typically demands more revolutionary changes, has a more radical rhetoric and ideology, and is less willing to compromise with political opponents.[92] With all that in mind, try reflecting on the climate action groups you know or are a part of. Where do you think the line is between the moderate and radical sides of the movement as a whole?

Both scholars of socio-ecological change and individuals involved in such change are continually trying to determine whether a radical flank has more positive or negative effects on social movements. One such scholar, Herbert Haines from the Western New England College, was the first to extensively research the impacts of radicals. Throughout his work studying the dynamics between radicals and mainstream activists within the civil rights movement, Haines developed the term "radical flank effect" to describe the impact a radical flank has on its overarching movement.[91,93] In his analysis of the radical flank effect of the 1960s civil rights movement, Haines determined, against popular opinion, that the presence of radicals positively affected the fundraising gains of moderate groups.[91]

While Haines found the radical flank of the civil rights movement had a positive effect, current research highlights the complexity of the radical flank as having both negative and positive effects. Notably, most of this research has used observational methods that do not allow causal claims. Let's look at what radical flanks can achieve for the worse or for the better.

Disadvantages of radical flanks

Studies have come to the conclusion that a radical flank has the potential to draw attention away from the actual aim of the collective action. In line with this, a recent study on the protests led by Black people between 1960 and 1972 found that more violent protests tended to steer public discourse towards the issue of "social control" and away from the issue of "civil rights".[94] We saw this effect play out with the Berlin protest actions of *Letzte Generation* [Last Generation], mentioned in Strategy 3. In the public discourse over their protests, the focus remained on decrying the actions and not the climate crisis, on prosecuting the protesters and not the perpetrators of climate injustice.

In the same vein, a radical flank can heighten the probability and degree of repression by the state.[95] This fits our example in Focus 2 – Strategy 4 of Chapter 2, which shows how protests can escalate when the police perceive a confrontational minority as representative of the crowd.

The actions of a radical flank within a given climate action group can further motivate people to vote for a party that opposes that climate action group. One study found that, between 1984 and 2012, the Green Party in the United

States received fewer electoral votes in districts in and nearby to which there had been forceful and violent pro-environmental protest actions than in districts in and nearby to which there had been none.[96] Similarly, violent protests led by Black people between 1960 and 1972 have likely influenced White people to vote for the Republicans.[94]

Other research indicates that the existence of a violent radical flank is associated with less participation in its non-violent moderate counterpart.[92] However, the direction of causality is not automatically a given here. Less participation in moderate groups could incite the creation of a radical flank. Notably, the same study found no general positive or negative effect of radical flanks on campaign success.[92]

Advantages of radical flanks

Radical flanks can also have many positive influences. Though it utilizes peaceful methods of protest, the Go Fossil Free divestment campaign is considered a radical flank within the climate movement because of its extreme rhetoric and extreme goal of achieving an "end [to] the age of fossil fuels"[97]. By analyzing emerging issues in 42,072 newspaper articles, researchers studying the Go Fossil Free campaign found that the radical flank had had the positive effect of bringing radical and liberal issues increasingly to the center of attention.[98]

A study analyzing the yearly progress of campaigns run between 1945 and 2006 found that a movement is more likely to advance its own political goals within a given year if a radical flank is present in that same year.[95] Several studies also showed that radical flanks can promote public support for political goals and moderate groups.[99–101] In a study on the Black Lives Matter protests in the summer of 2020, researchers found that when there was a mix of non-violent and violent protests, support for the movement's policy goals increased among conservatives living in relatively liberal areas.[99] And an experimental study comparing various newspaper articles on an animal rights movement and on the climate movement found causal evidence for the presence of a radical flank leading to more support for the moderate group within the movement.[100] It is worth noting, however, that while support for moderate groups may increase, research also suggests that the contrast between moderates and radicals can also lead to a decrease in support for the radical flank itself.[101]

Finding the right balance

These mixed and sometimes ambiguous findings highlight how difficult it can be to carry out a collective action that finds the right balance between public attention and support. Since collective actions don't exist in a vacuum, what we can learn from the radical flank effects described in studies is that each action group operates in a vast network of actors within their movement, with each of these groups and actors having the ability to influence each other's success. Every action that one climate action group takes, whether more or less radical, can affect the public's perception of other groups participating in different collective actions or the overarching movement.

With that in mind, it might be worthwhile to reflect on the aims of each climate action group within the climate movement and then continually monitor the impact of their actions on the greater whole. Seeking discussions and collaborations with allied groups may support this process.

 Box 3.9: The bottom line

Evidence shows that a radical flank can have both positive and negative impacts within a given movement. A radical flank may draw attention away from the actual goals, decrease votes for parties that align with the cause, and potentially hamper the mobilizing efforts of more moderate groups. At the same time, a radical flank may bring radical ideas to the center of attention and increase support for their political goals and moderate counterparts. A key takeaway from studies on the radical flank effect is that the collective action practices of one climate action group inevitably affect those of other groups.

Focus 2 – Strategy 5: Using constructive disruption to reduce resistance to climate goals

One particular method that could be effective in gaining attention *and* public support – not necessarily for the group itself but for its political goals – is creating constructive disruption.

A study by Eric Shuman and colleagues from the University of Groningen, the Hebrew University of Jerusalem, and the Interdisciplinary Center Herzliya examined three types of collective action and their effect on people initially resistant to supporting the political demands of the portrayed social movements.[102] These actions were: normative non-violent actions (peaceful demonstrations, petitions), non-normative non-violent actions (strikes, road blockages, sit-ins), and non-normative violent actions (riots, property destruction). Across different social movements (not including any climate movements), this study found that resistant individuals were particularly willing to support political concessions if the collective actions were non-normative and non-violent. Shuman and colleagues labeled the dynamic created by this type of collective action "constructive disruption".

Within the context of collective action, a "disruption" refers to an action that cannot be ignored. One extreme example of disruptive action is an openly violent rebellion. Examples of less extreme disruptive action include street blockades or acts of sabotage like damaging a pipeline. The authors of this study argue that an action with a lower level of disruption is more easily ignored by people initially disinclined to support a social movement's goals.[102] This idea tells us that a certain level of disruption can be seen as beneficial and may even be necessary.

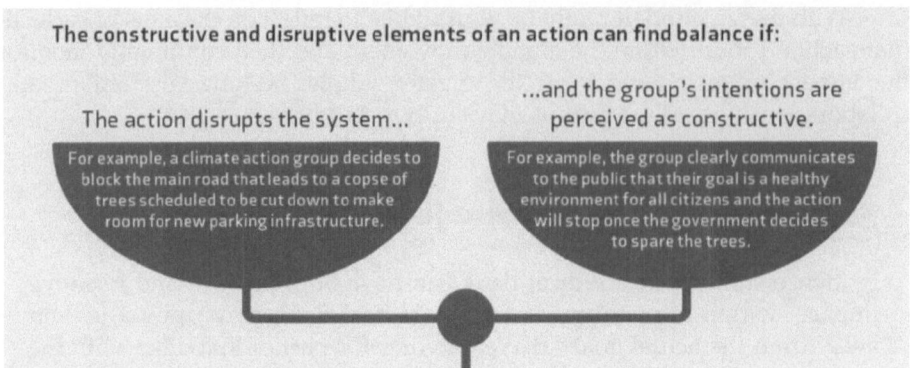

The constructive and disruptive elements of an action can find balance if:

The action disrupts the system...

For example, a climate action group decides to block the main road that leads to a copse of trees scheduled to be cut down to make room for new parking infrastructure.

...and the group's intentions are perceived as constructive.

For example, the group clearly communicates to the public that their goal is a healthy environment for all citizens and the action will stop once the government decides to spare the trees.

Figure 3.3: A balanced disruptive and constructive collective climate action

However, for maximum effectiveness, an action's disruptive elements must be balanced with constructive intentions. Regardless of how disruptive an action is, outsiders must perceive the people involved as acting constructively. Within the context of collective action, a "constructive" action is one in which outsiders believe that those involved are willing to desist once their demands are met and that those involved are not creating disruption for the sole purpose of being destructive but out of a true desire to seek an improvement in the situation for everyone.[102] An example of a constructive disruption can be seen in Figure 3.3.

As can be seen in Figure 3.3, to be perceived as a group driven by constructive intentions, it is beneficial to engage in actions with clear rationales. Put simply: opt for protest actions that speak for themselves. For example, it is intuitively easier to understand why climate action groups might block coal power plants than why they might block public transportation. The first action has a clear symbolic effect (as well as the direct effect of blocking the production of greenhouse gases). The second action, on the other hand, can easily be perceived as unconstructive.

The climate action group *Extinction Rebellion* had to deal with this kind of undesirable effect when, in 2019, certain group members blocked parts of the London metro system during rush hour[103,104]. Many commuters reacted very angrily and even violently towards the participating activists. It later came out that many members of *Extinction Rebellion* did not support this choice of action, with some denouncing the action as a mistake. The group announced they would learn from the event, re-evaluate their internal decision-making processes,[103] and move away from this type of disruption.

It is worth noting that the perceived constructiveness of an action is highly context-dependent. In one context, it may be considered appropriate to shut down an entire metro system for a day, and in another context, it may seem entirely inappropriate. Even though *Extinction Rebellion* did achieve attention with their choice of disruptive action, they did not achieve the desired balance of disruption and constructive intention within the public perception. However, a recent study suggests that despite the negative reactions to this disruptive action *Extinction Rebellion*'s work overall still had a positive effect on reducing opposition to pro-environmental policies.[105]

 Box 3.10: The bottom line

If a disruptive action can be seen as fueled by constructive intentions, it can generate support for the acting group's political goals, even among those who are initially resistant to the group's advocated changes. Non-normative non-violent action, such as civil disobedience, has the greatest propensity for achieving the balance between disruption and constructive intentions, by demonstrating to outsiders that the action is aimed at improving the situation for all and will desist once the group's goals are met.

Focus 2 – Strategy 6: Ensuring that a climate action is seen as legitimate, relatable, and effective

Not everyone will be ready or willing to engage in disruptive action. And even for those who are and see themselves as having constructive intentions, the public can still wind up seeing these types of protests as nothing more than destructive. Indeed, even the media coverage of a largely peaceful protest can end up coloring the action as a whole as immoral or dangerous, thereby negatively impacting public support. This was precisely the case with the 2017 G20 protests in Hamburg. During these otherwise peaceful protests, a small group of activists engaged in acts of vandalism. As a result, all of the protests received negative press, and the focus was drawn away from what was being protested to the violence and criminality of the protesters themselves, thus shaping the public's perception of the collective action in its entirety.[106]

What examples like these tell us is that in addition to being selective about which types of collective action we participate in (those more likely to be perceived as constructive), it is also important for climate action groups to design actions that will be perceived as legitimate, relatable, and effective in achieving change.

Ensuring that a climate action is seen as legitimate

Ensuring a collective action is perceived as legitimate and reasonable is particularly important when that collective action might also be perceived as controversial and non-normative. We've already learned how the police's perception of the legitimacy of a protest can influence whether it escalates (see Focus 1 – Strategy 4 in Chapter 2 for more on how peaceful protests escalate).

With respect to the general public, collective actions that are perceived as legitimate, reasonable, and moral generate more support among those not involved in the movement.[71,85–87] And, one method of creating an air of legitimacy, reasonableness, and morality around an action is to clearly communicate its justification – by highlighting the moral injustice against which people are protesting and explaining why that specific form of collective action was chosen. An example of this method might look like, "we want to stop this injustice directly where it's happening", or "we've tried everything else, and nothing has worked, so this is our last resort". What's most important here is that there is consensus among the members of a given group on what communication strategy is going to be used to justify their actions – ideally a strategy that keeps the values of the target audience in mind.

Another method for this kind of communication is highlighting the level of corruption within a given context (e.g., whale hunting), so that an action is more easily perceived as justified and legitimate.[86] An impressive example of this can be seen in the 2016–2017 disruptive action on the Dakota Access Pipeline. Ruby Montoya and Jessica Reznicek had tried all legal means in their attempt to stop the pipeline's construction – they gathered signatures, attended public meetings, and joined demonstrations – with little success.[107] Frustrated by the lack of progress, they decided to take matters into their own hands. On one night in November of 2016, these two women set fire to Dakota Access Pipeline equipment to the tune of $2.5 million in damages. Over the following six months, they continued setting fire to integral pipeline components. But their actions went largely unreported by the media. And the pipeline continued to be built. There was only one thing the two women felt they had left to try in order to bring substantial attention to the issue. Montoya and Reznicek drove to the responsible state regulatory agency. Standing next to the agency's sign, they publicly confessed their acts of eco-sabotage to the media.[107] They stated that while some might perceive their acts as violent, they saw it as necessary resistance against a private corporation that polluted water and seized land. They emphasized that they never harmed human life but only acted out of deep empathy. Ruby Montoya and Jessica Reznicek now face up to 20 years in prison – a sentence that would be among the longest ever served for eco-activism in the U.S. *Can you empathize with these two activists? Does their reasoning make you more inclined to describe the disruptive action as legitimate?*

Ensuring that a climate action is seen as relatable and effective

How likely it is that we perceive an action as legitimate and how willing we are to support a movement or group is affected by our ability to identify with it.[85] People from outside as well as people inside a movement may perceive a non-normative collective action as immoral, which makes them less likely to identify with the action or the group. Moreover, the less people identify with an action, the less likely they are to support the action's overarching movement or the goals of that movement.[71]

In fact, studies on radical flanks found that social identification may play an important role in a radical flank's effect on public support. On the one hand, when people are instructed to compare a moderate group with a radical flank, this can decrease identification with, and therefore support for, the radical flank. On the other, seeing these differences can also provide the basis for identifying and supporting more moderate groups.[100,101] Therefore, when planning a collective action, climate action groups, and radical flanks in particular, might benefit from reflecting on how best to create a basis for identification (see Chapter 2 for more on social identification).

Research has also found that outsiders report being more willing to join future protests if they see a group's chosen tactics as logical and effective at achieving change.[87] Interestingly, this research also indicates that normative peaceful demonstrations were perceived as the most legitimate but not the most effective. Instead, non-normative peaceful protests (constructive disruptions) were viewed as more effective. This perception of effectiveness is closely tied to a group's efficacy beliefs, which we'll learn more about in Chapter 5.

The key takeaway here is that when it comes to designing actions, climate action groups would do well to put effort into clearly communicating to outsiders and members alike why a particular tactic is being chosen and how that tactic will be effective in promoting socio-ecological change.

 Box 3.11: The bottom line

Normative collective actions receive public support more easily because they are more likely to be perceived as legitimate. People engaged in non-normative collective action, on the other hand, need to make an effort to convince others that their actions are legitimate and morally justified if they want future support. A group that presents itself and its actions as relatable and effective can build support among those outside the movement.

DISCOVERING YOUR VALUE BASIS

Reading through this chapter on beliefs and emotions might have felt somewhat intense. So, it might by a good idea to take some time now to reflect on the values, moral beliefs, and core principles that guide you in your life. To that end, take a look at the following activity, which has been divided into two steps and is designed to help guide your reflection process.

Before getting started, it might be helpful to check out Figure 3.4 for an example of roughly how this task should look.

Step 1: Forming an overview of your value basis

1. Draw a circle that has 7 concentric circles inside it. Label each circle 1 through 7, ascending numerically to the outermost ring. Divide the circle up into sections that each represent the values presented in this chapter. Alternatively, you can download a template from the *Wandelwerk* website[108] or work through the task in your head.
2. For each value, make a dot on the dividing line that aligns with the level of importance that value has for you, with 7 being extremely important and 1 being unimportant.
3. Connect the dots to get a clearer picture of your value basis.

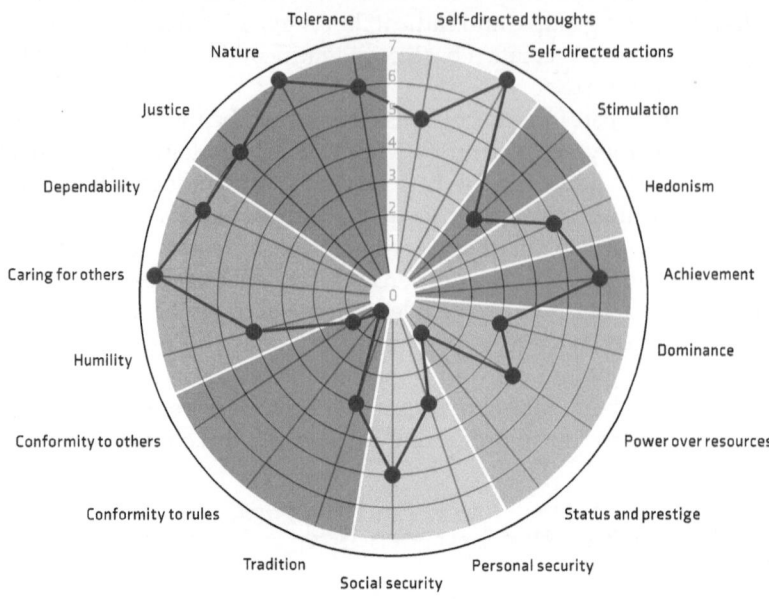

Figure 3.4: Example of a completed value wheel

Step 2: Reflecting on your moral beliefs and values

1. Find a different colored pen. For each value, now make a dot on the dividing line that aligns with the level of action you currently take towards pursuing that value in your life, with 7 meaning it is heavily pursued and 1 meaning it is not pursued at all.
2. For each value, how far apart or close together are these dots? Perhaps you assigned high importance (7) to the value of "justice" but determined you don't actually do very much to actively pursue justice (2). Or perhaps you assigned low importance (1) to the value of "dominance" but you pursued a career that requires you to take dominant positions (4).
3. Take a moment to assess your dots. Once you've done that, make your way through the following questions:
   - *How does the overall difference between my values and my current life look?*
   - *Is there anything that surprises me?*
   - *Is there anything that bothers me or threatens my self-image?*
4. Determine your moral beliefs by asking yourself:
   - *Which values have a clear right or wrong for me?*
   - *Which values would make me angry if they were violated?*
5. Determine how your values align with your social network by asking yourself:
   - *Which of the values that are important to me do I express in conversations with my family members, friends, and colleagues?*
   - *Do I feel like they understand my moral beliefs?*
   - *Do I understand their moral beliefs?*

6. Determine your place in collective climate action by asking yourself:
   - *Am I involved in any particular collective climate actions?*
   - *If so, how do they reflect my values?*
   - *If not, what collective actions could I imagine might align with my values?*

References

1. Nakate, V. & Thunberg, G. An open letter to the global media by Greta Thunberg and Vanessa Nakate. *TIME.* https://time.com/6111851/greta-thunberg-vanessa-nakate-open-letter-media/ (2021).
2. Skitka, L. J. & Bauman, C. W. Moral conviction and political engagement: moral conviction. *Polit. Psychol.* 29, 29–54 (2008). https://doi.org/10.1111/j.1467-9221.2007.00611.x
3. Pearson, A. R., Tsai, C. G. & Clayton, S. Ethics, morality, and the psychology of climate justice. *Curr. Opin. Psychol.* 42, 36–42 (2021). https://doi.org/10.1016/j.copsyc.2021.03.001
4. Kleres, J. & Wettergren, Å. Fear, hope, anger, and guilt in climate activism. *Soc. Mov. Stud.* 16, 507–519 (2017). https://doi.org/10.1080/14742837.2017.1344546
5. Schwartz, S. H. Basic human values: An overview. *Hebr. Univ. Jerus.* http://www.yourmorals.org/schwartz (2006).
6. Schwartz, S. H. Universals in the content and structure of values: Theoretical advances and empirical tests in 20 countries. *Adv. Exp. Soc. Psychol.* 25, 1–65 (1992). https://doi.org/10.1016/S0065-2601(08)60281-6
7. Schwartz, S. H. *et al.* Value tradeoffs propel and inhibit behavior: Validating the 19 refined values in four countries. *Eur. J. Soc. Psychol.* 47, 241–258 (2017). https://doi.org/10.1002/ejsp.2228
8. Steg, L. & de Groot, J. I. M. Environmental Values. In *The Oxford Handbook of Environmental and Conservation Psychology* (ed. Clayton, S. D.) 81–92 (Oxford University Press, 2012). https://doi.org/10.1093/oxfordhb/9780199733026.013.0005
9. Almers, E. Pathways to action competence for sustainability – Six themes. *J. Environ. Educ.* 44, 116–127 (2013). https://doi.org/10.1080/00958964.2012.719939
10. Chawla, L. Significant life experiences revisited: A review of research on sources of environmental sensitivity. *J. Environ. Educ.* 29, 11–21 (1998). https://doi.org/10.1080/00958969809599114
11. Fisher, S. R. Life trajectories of youth committing to climate activism. *Environ. Educ. Res.* 22, 229–247 (2016). https://doi.org/10.1080/13504622.2015.1007337
12. Sheldon, K. M., Wineland, A., Venhoeven, L. & Osin, E. Understanding the motivation of environmental activists: A comparison of self-determination theory and functional motives theory. *Ecopsychology* 8, 228–238 (2016). https://doi.org/10.1089/eco.2016.0017
13. Friends of the Earth. Homepage. *Friends of the Earth.* https://foe.org/
14. Transition Network. Transition Towns. https://transitionnetwork.org/ (2016).
15. weit. Film – Weit um die Welt. https://www.weitumdiewelt.de/film/
16. Solar Butterfly. Start. https://solarbutterfly.org/
17. Lampert, M., Inglehart, R., Metaal, S., Schoemaker, H. & Papadongonas, P. COVID-19 crisis boosts progressive values amidst growing pessimism – Measuring the pandemic's impact on social values, emotions and priorities in 24 countries. *Glocalities* 1–46 (2021).
18. Agostini, M. & van Zomeren, M. Toward a comprehensive and potentially cross-cultural model of why people engage in collective action: A quantitative research

synthesis of four motivations and structural constraints. *Psychol. Bull.* 147, 667–700 (2021). https://doi.org/10.1037/bul0000256

19. van Zomeren, M., Pauls, I. L. & Cohen-Chen, S. Is hope good for motivating collective action in the context of climate change? Differentiating hope's emotion- and problem-focused coping functions. *Glob. Environ. Change* 58, 101915 (2019). https://doi.org/10.1016/j.gloenvcha.2019.04.003

20. Nguyen, Q. N., Nguyen, D. M. & Nguyen, L.V. An examination of the social identity model of collective action in the context of Vietnam. *Open Psychol. J.* 14, 1–10 (2021). https://doi.org/10.2174/1874350102114010001

21. Fernandes-Jesus, M., Lima, M. L. & Sabucedo, J.-M. "Save the climate! Stop the oil": Actual protest behavior and core framing tasks in the Portuguese climate movement. *J. Soc. Polit. Psychol.* 8, 426–452 (2020). https://doi.org/10.5964/jspp.v8i1.1116

22. Brügger, A., Gubler, M., Steentjes, K. & Capstick, S. B. Social identity and risk perception explain participation in the swiss youth climate strikes. *Sustainability* 12, 10605 (2020). https://doi.org/10.3390/su122410605

23. Marczak, M., Winkowska, M., Chaton-Østlie, K., Morote Rios, R., & Klöckner, C. A. "When I say I'm depressed, it's like anger": An exploration of the emotional landscape of climate change concern in Norway and its psychological, social and political implications. *Emotion, Space and Society*, 46, 100939 (2023). https://doi.org/10.21203/rs.3.rs-224032/v2

24. Thomas, E. F., McGarty, C. & Mavor, K. I. Aligning identities, emotions, and beliefs to create commitment to sustainable social and political action. *Personal. Soc. Psychol. Rev.* 13, 194–218 (2009). https://doi.org/10.1177/1088868309341563

25. Hofmann, W., Wisneski, D. C., Brandt, M. J. & Skitka, L. J. Morality in everyday life. *Science* 345, 1340–1343 (2014). https://doi.org/10.1126/science.1251560

26. Kutlaca, M., van Zomeren, M. & Epstude, K. Our right to a steady ground: Perceived rights violations motivate collective action against human-caused earthquakes. *Environ. Behav.* 51, 315–344 (2017). https://doi.org/10.1177/0013916517747658

27. Mazzoni, D., van Zomeren, M. & Cicognani, E. The motivating role of perceived right violation and efficacy beliefs in identification with the Italian water movement: Explaining water activism. *Polit. Psychol.* 36, 315–330 (2015). https://doi.org/10.1111/pops.12101

28. Keshavarzi, S., McGarty, C. & Khajehnoori, B. Testing social identity models of collective action in an Iranian environmental movement. *J. Community Appl. Soc. Psychol.* 31, 452–464 (2021). https://doi.org/10.1002/casp.2523

29. Pauls, I. L., Shuman, E., van Zomeren, M., Saguy, T. & Halperin, E. Does crossing a moral line justify collective means? Explaining how a perceived moral violation triggers normative and nonnormative forms of collective action. *Eur. J. Soc. Psychol.* 52, 105–123 (2022). https://doi.org/10.1002/ejsp.2818

30. van Zomeren, M., Kutlaca, M. & Turner-Zwinkels, F. Integrating who "we" are with what "we" (will not) stand for: A further extension of the *Social Identity Model of Collective Action. Eur. Rev. Soc. Psychol.* 29, 122–160 (2018). https://doi.org/10.1080/10463283.2018.1479347

31. Yard Digital PR Team. Just plane wrong: Celebs with the worst private jet Co2 emissions. *Yard.* https://weareyard.com/insights/worst-celebrity-private-jet-co2-emission-offenders (2022).

32. We move Europe. Ban private jets! https://act.wemove.eu/campaigns/ban-private-jets/

33. van Zomeren, M., Postmes, T. & Spears, R. Toward an integrative social identity model of collective action: A quantitative research synthesis of three socio-psychological

perspectives. *Psychol. Bull.* 134, 504–535 (2008). https://doi.org/10.1037/0033-2909.134.4.504

34. Jasper, J. M. *The Emotions of Protest*. (The University of Chicago Press, 2018). https://doi.org/10.7208/chicago/9780226561813.001.0001

35. Skitka, L. J. The psychology of moral conviction. *Soc. Personal. Psychol. Compass* 4, 267–281 (2010). https://doi.org/10.1111/j.1751-9004.2010.00254.x

36. Haidt, J. The Moral Emotions. In *Handbook of Affective Sciences* (eds. Davidson, R. J., Scherer, K. R. & Goldsmith, H. H.) 852–870 (Oxford University Press, 2003). https://doi.org/10.1093/oso/9780195126013.003.0045

37. Harth, N. S., Leach, C. W. & Kessler, T. Guilt, anger, and pride about in-group environmental behaviour: Different emotions predict distinct intentions. *J. Environ. Psychol.* 34, 18–26 (2013). https://doi.org/10.1016/j.jenvp.2012.12.005

38. Ferguson, M. A. & Branscombe, N. R. Collective guilt mediates the effect of beliefs about global warming on willingness to engage in mitigation behavior. *J. Environ. Psychol.* 30, 135–142 (2010). https://doi.org/10.1016/j.jenvp.2009.11.010

39. Rees, J. H. & Bamberg, S. Climate protection needs societal change: Determinants of intention to participate in collective climate action. *Eur. J. Soc. Psychol.* 44, 466–473 (2014). https://doi.org/10.1002/ejsp.2032

40. Haugestad, C. A. P., Skauge, A. D., Kunst, J. R. & Power, S. A. Why do youth participate in climate activism? A mixed-methods investigation of the #FridaysForFuture climate protests. *J. Environ. Psychol.* 76, 101647 (2021). https://doi.org/10.1016/j.jenvp.2021.101647

41. Ekman, P. An argument for basic emotions. *Cogn. Emot.* 6, 169–200 (1992). https://doi.org/10.1080/02699939208411068

42. Bamberg, S., Rees, J. & Seebauer, S. Collective climate action: Determinants of participation intention in community-based pro-environmental initiatives. *J. Environ. Psychol.* 43, 155–165 (2015). https://doi.org/10.1016/j.jenvp.2015.06.006

43. Hickman, C. *et al.* Climate anxiety in children and young people and their beliefs about government responses to climate change: A global survey. *Lancet Planet. Health* 5, e863–e873 (2021). https://doi.org/10.1016/S2542-5196(21)00278-3

44. Bührle, H. & Kimmerle, J. Psychological determinants of collective action for climate justice: Insights from semi-structured interviews and content analysis. *Front. Psychol.* 12, 695365 (2021). https://doi.org/10.3389/fpsyg.2021.695365

45. Stanley, S. K., Hogg, T. L., Leviston, Z. & Walker, I. From anger to action: Differential impacts of eco-anxiety, eco-depression, and eco-anger on climate action and well-being. *J. Clim. Change Health* 1, 100003 (2021). https://doi.org/10.1016/j.joclim.2021.100003

46. Wallis, H. & Loy, L. S. What drives pro-environmental activism of young people? A survey study on the Fridays For Future movement. *J. Environ. Psychol.* 74, 101581 (2021). https://doi.org/10.1016/j.jenvp.2021.101581

47. Furlong, C. & Vignoles, V. L. Social identification in collective climate activism: Predicting participation in the environmental movement, extinction rebellion. *Identity* 21, 20–35 (2021). https://doi.org/10.1080/15283488.2020.1856664

48. Gregersen, T., Andersen, G. & Tvinnereim, E. The strength and content of climate anger. *Glob. Environ. Change* 82, 102738 (2023). https://doi.org/10.1016/j.gloenvcha.2023.102738

49. Landmann, H. & Rohmann, A. Being moved by protest: Collective efficacy beliefs and injustice appraisals enhance collective action intentions for forest protection via positive and negative emotions. *J. Environ. Psychol.* 71, 101491 (2020). https://doi.org/10.1016/j.jenvp.2020.101491

50. Landmann, H. & Naumann, J. Being positively moved by climate protest predicts peaceful collective action. *Glob. Environ. Psychol.* https://www.psycharchives.org/en/item/72fb35c7-7174-47b6-985c-7d1a477965db (2023).

51. Sabherwal, A. *et al.* Anger consensus messaging can enhance expectations for collective action and support for climate mitigation. *J. Environ. Psychol.* 76, 101640 (2021). https://doi.org/10.1016/j.jenvp.2021.101640

52. Hamann, K. R. S., Dasch, S. T., & von Agris, A.-S. Environmental Collective Action in Germany and Beyond – An Opportunity to Extend Theory and Practice. In *The Social and Political Psychology of Protest across and within Cultures* (ed. van Zomeren, M.) (Routledge, 2025).

53. de Moor, J., Uba, K., Wahlström, M., Wennerhag, M. & De Vydt, M. *Protest for a Future II – Composition, Mobilization and Motives of the Participants in Fridays For Future Climate Protests on 20-27 September, 2019, in 19 Cities around the World.* https://doi.org/10.17605/osf.io/asruw (2020).

54. du Bray, M., Wutich, A., Larson, K. L., White, D. D. & Brewis, A. Anger and sadness: Gendered emotional responses to climate threats in four island nations. *Cross-Cult. Res.* 53, 58–86 (2019). https://doi.org/10.1177/1069397118759252

55. Plant, E. A., Hyde, J. S., Keltner, D. & Devine, P. G. The gender stereotyping of emotions. *Psychol. Women Q.* 24, 81–92 (2000). https://doi.org/10.1111/j.1471-6402.2000.tb01024.x

56. Brescoll, V. L. & Uhlmann, E. L. Can an angry woman get ahead?: Status conferral, gender, and expression of emotion in the workplace. *Psychol. Sci.* 19, 268–275 (2008). https://doi.org/10.1111/j.1467-9280.2008.02079.x

57. Democracy Now. "We are striking to disrupt the system": An hour with 16-year-old climate activist Greta Thunberg. https://www.democracynow.org/2019/9/11/greta_thunberg_swedish_activist_climate_crisis (2019).

58. Global Coal Exit List. GCEL 2023. https://www.coalexit.org/

59. Smoltczyk, A. Maldives president leads the charge against climate change. *SPIEGEL international.* https://www.spiegel.de/international/world/the-underwater-obama-maldives-president-leads-the-charge-against-climate-change-a-658373.html (2009).

60. Thomas, E. F. & McGarty, C. A. The role of efficacy and moral outrage norms in creating the potential for international development activism through group-based interaction. *Br. J. Soc. Psychol.* 48, 115–134 (2009). https://doi.org/10.1348/014466608X313774

61. Thomas, E. F., McGarty, C. & Mavor, K. Group interaction as the crucible of social identity formation: A glimpse at the foundations of social identities for collective action. *Group Process. Intergroup Relat.* 19, 137–151 (2016). https://doi.org/10.1177/1368430215612217

62. BMUV & UBA. Umweltbewusstsein in Deutschland 2022 – Ergebnisse einer repräsentativen Bevölkerungsumfrage. (2022).

63. Hamann, K. R. S., von Agris, A.-S. & Markus, L. Investigating the predictors of collective action intensity and health. https://osf.io/preprints/psyarxiv/qev28_v1 (2023).

64. Debt for Climate. Home. *Debtforclimate.* https://www.debtforclimate.org

65. Debt for Climate! Debt for Climate Global Mobilization Oct–Nov 2022. https://www.youtube.com/watch?v=JlvrnNg7usI (2023).

66. Burstein, P. The impact of public opinion on public policy: A review and an agenda. *Polit. Res. Q.* 56, 29–40 (2003). https://doi.org/10.1177/106591290305600103

67. Burstein, P. & Linton, A. The impact of political parties, interest groups, and social movement organizations on public policy: Some recent evidence and theoretical concerns. *Soc. Forces* 81, 380–408 (2002). https://doi.org/10.1353/sof.2003.0004

68. Louis, W. R. Collective action – And then what? *J. Soc. Issues* 65, 727–748 (2009). https://doi.org/10.1111/j.1540-4560.2009.01623.x

69. Muñoz, J. & Anduiza, E. 'If a fight starts, watch the crowd': The effect of violence on popular support for social movements. *J. Peace Res.* 56, 485–498 (2019). https://doi.org/10.1177/0022343318820575

70. Chenoweth, E. & Stephan, M. J. *Why Civil Resistance Works: The Strategic Logic of Nonviolent Conflict.* (Columbia University Press, 2011). https://doi.org/10.17813/1086-671X-20-4-427

71. Feinberg, M., Willer, R. & Kovacheff, C. The activist's dilemma: Extreme protest actions reduce popular support for social movements. *J. Pers. Soc. Psychol.* 119, 1086–1111 (2020). https://doi.org/10.1037/pspi0000230

72. Lockhart, I. & Kramer, K. Deutschland spricht 2021: Ergebnisse zur Debattenaktion. https://www.faz.net/aktuell/deutschland-spricht/deutschland-spricht-2021-erg ebnisse-zur-debattenaktion-17570629.html (2021).

73. Ksenofontov, I. & Becker, J. C. The harmful side of thanks: Thankful responses to high-power group help undermine low-power groups' protest. *Pers. Soc. Psychol. Bull.* 46, 794–807 (2020). https://doi.org/10.1177/0146167219879125

74. Ellemers, N., van der Toorn, J., Paunov, Y. & van Leeuwen, T. The psychology of morality: A review and analysis of empirical studies published from 1940 through 2017. *Personal. Soc. Psychol. Rev.* 23, 332–366 (2019). https://doi.org/10.1177/10888 68318811759

75. Monin, B., Sawyer, P. J. & Marquez, M. J. The rejection of moral rebels: Resenting those who do the right thing. *J. Pers. Soc. Psychol.* 95, 76–93 (2008). https://doi.org/10.1037/0022-3514.95.1.76

76. Brouwer, C., Bolderdijk, J., Cornelissen, G. & Kurz, T. Communication strategies for moral rebels: How to talk about change in order to inspire self-efficacy in others. *WIREs Clim. Change* 13, (2022). https://doi.org/10.1002/wcc.781

77. Truelove, H. B., Carrico, A. R., Weber, E. U., Raimi, K. T. & Vandenbergh, M. P. Positive and negative spillover of pro-environmental behavior: An integrative review and theoretical framework. *Glob. Environ. Change* 29, 127–138 (2014). https://doi.org/10.1016/j.gloenvcha.2014.09.004

78. Rosenberg, M. B. & Chopra, D. *Nonviolent Communication: A Language of Life.* (PuddleDancer Press, 2015).

79. Wacker, R. & Dziobek, I. Preventing empathic distress and social stressors at work through nonviolent communication training: A field study with health professionals. *J. Occup. Health Psychol.* 23, 141–150 (2018). https://doi.org/10.1037/ocp0000058

80. Museux, A.-C., Dumont, S., Careau, E. & Milot, É. Improving interprofessional col-laboration: The effect of training in nonviolent communication. *Soc. Work Health Care* 55, 427–439 (2016). https://doi.org/10.1080/00981389.2016.1164270

81. Sung, J. & Kweon, Y. Effects of a nonviolent communication-based empathy educa-tion program for nursing students: A quasi-experimental pilot study. *Nurs. Rep.* 12, 824–835 (2022). https://doi.org/10.3390/nursrep12040080

82. Forschungsgruppe Wahlen e.V. Politbarometer: Wichtige Probleme in Deutschland. https://www.forschungsgruppe.de/Umfragen/Politbarometer/Langzeitentwicklu ng_-_Themen_im_Ueberblick/Politik_II/ (2023).

83. Russew, G.-S. & Witzki, V. Unterstützer der 'Letzten Generation' attackieren Monet-Gemälde. *rbb24.* https://www.rbb24.de/panorama/beitrag/2022/10/brandenb urg-barberini-letzte-generation-bild-attackiert.html (2022).

84. Pleul, P. Nach Aktion auf Flughafen BER: 'Letzte Generation' im Kreuzfeuer der Kritik. *Stern.* https://www.stern.de/politik/deutschland/-letzte-generation---mass ive-kritik-nach-blockade-am-flughafen-ber-32947098.html (2022).

85. Simpson, B., Willer, R. & Feinberg, M. Does violent protest backfire? Testing a theory of public reactions to activist violence. *Socius Sociol. Res. Dyn. World* 4, 237802311880318 (2018). https://doi.org/10.1177/2378023118803189

86. Thomas, E. F. & Louis, W. R. When will collective action be effective? Violent and non-violent protests differentially influence perceptions of legitimacy and efficacy among sympathizers. *Pers. Soc. Psychol. Bull.* 40, 263–276 (2014). https://doi.org/10.1177/0146167213510525

87. Zlobina, A. & Gonzalez Vazquez, A. What is the right way to protest? On the process of justification of protest, and its relationship to the propensity to participate in different types of protest. *Soc. Mov. Stud.* 17, 234–250 (2018). https://doi.org/10.1080/14742837.2017.1393408

88. Gamson, W. A. *The Strategy of Social Protest*. (Homewood, IL: Dorsey Press, 1975).

89. Myers, D. J. & Caniglia, B. S. All the rioting that's fit to print: Selection Effects in national newspaper coverage of civil disorders, 1968–1969. *Am. Sociol. Rev.* 69, 519–543 (2004). https://doi.org/10.1177/000312240406900403

90. Sobieraj, S. Reporting conventions: Journalists, activists, and the thorny struggle for political visibility. *Soc. Probl.* 57, 505–528 (2010). https://doi.org/10.1525/sp.2010.57.4.505

91. Haines, H. H. *Black Radicals and the Civil Rights Mainstream, 1954–1970*. (University of Tennessee Press, 1988). https://doi.org/10.1086/ahr/95.4.1320-a

92. Chenoweth, E. & Schock, K. Do contemporaneous armed challenges affect the outcomes of mass nonviolent campaigns? *Mobilization Int. Q.* 20, 427–451 (2015). https://doi.org/10.17813/1086-671X-20-4-427

93. Haines, H. H. Radical Flank Effects. In *The Wiley-Blackwell Encyclopedia of Social and Political Movements* (eds. Snow, D. A., Della Porta, D., Klandermans, B. & McAdam, D.) wbespm174 (Blackwell Publishing Ltd, 2013). https://doi.org/10.1002/9780470674871.wbespm174

94. Wasow, O. Agenda seeding: How 1960s black protests moved elites, public opinion and voting. *Am. Polit. Sci. Rev.* 114, 638–659 (2020). https://doi.org/10.1017/S000305542000009X

95. Tompkins, E. A Quantitative Reevaluation of Radical Flank Effects within Nonviolent Campaigns. In *Research in Social Movements, Conflicts and Change* (ed. Coy, P. G.) vol. 38 103–135 (Emerald Group Publishing Limited, 2015). https://doi.org/10.1108/S0163-786X20150000038004

96. Farrer, B. & Klein, G. R. How radical environmental sabotage impacts US elections. *Terror. Polit. Violence* 34, 218–239 (2019). https://doi.org/10.1080/09546553.2019.1678468

97. Fossil Free. We can build a fossil free world. https://gofossilfree.org/

98. Schifeling, T. & Hoffman, A. J. Bill McKibben's influence on U.S. climate change discourse: Shifting field-level debates through radical flank effects. *Organ. Environ.* 32, 213–233 (2019). https://doi.org/10.1177/1086026617744278

99. Shuman, E., Hasan-Aslih, S., Van Zomeren, M., Saguy, T. & Halperin, E. Protest movements involving limited violence can sometimes be effective: Evidence from the 2020 BlackLivesMatter protests. *Proc. Natl. Acad. Sci.* 119, e2118990119 (2022). https://doi.org/10.1073/pnas.2118990119

100. Simpson, B., Willer, R. & Feinberg, M. Radical flanks of social movements can increase support for moderate factions. *PNAS Nexus* 1, pgac110 (2022). https://doi.org/10.1093/pnasnexus/pgac110

101. Dasch, S., Bellm, M., Shuman, E. & van Zomeren, M. The radical flank: Curse or blessing of a social movement? *Glob. Environ. Psychol.* https://www.psycharchives.org/en/item/7cf21891-1dd0-4c98-9bf2-faf9142dca42 (2023).

102. Shuman, E., Saguy, T., van Zomeren, M. & Halperin, E. Disrupting the system constructively: Testing the effectiveness of nonnormative nonviolent collective action. *J. Pers. Soc. Psychol.* 121, 819–841 (2021). https://doi.org/10.1037/pspi0000333

103. Townsend, M. Tube protest was a mistake, admit leading Extinction Rebellion members. *The Guardian.* https://www.theguardian.com/environment/2019/oct/20/extinction-rebellion-tube-protest-was-a-mistake (2019).

104. Gayle, D. & Quinn, B. Extinction Rebellion rush-hour protest sparks clash on London Underground. *The Guardian.* https://www.theguardian.com/environment/2019/oct/17/extinction-rebellion-activists-london-underground (2019).

105. Kountouris, Y. & Williams, E. Do protests influence environmental attitudes? Evidence from Extinction Rebellion. *Environ. Res. Commun.* 5, 011003 (2023). https://doi.org/10.1088/2515-7620/ac9aeb

106. Biermann, K., Geisler, A. & Polke-Majewski, K. Krawalle beim G20-Gipfel: Wer waren die Gewalttäter von Hamburg? *Die Zeit.* https://www.zeit.de/politik/deutschland/2017-07/krawalle-g20-gipfel-hamburg-gewalttaeter (2017).

107. Shipley, J. 'You strike a match': Why two women sacrificed everything to stop the Dakota Access Pipeline. *Roll. Stone.* https://www.rollingstone.com/culture/culture-features/dakota-access-pipeline-eco-sabotage-jessica-reznicek-ruby-montoya-1173735/ (2021).

108. Wandelwerk e.V. Wandelwerk Umweltpsychologie: Wir bringen Psychologie in den Umweltschutz. https://www.wandel-werk.org/

4 FRAMING CLIMATE ACTION

DEFINING FRAMING

Across various groups, including the field of climate communication, "framing" has become an increasingly popular buzzword. As, within this book, the idea has already been introduced that it may be a source of motivation to highlight specific aspects of a collective climate action (such as its legitimacy, relatability, and

DOI: 10.4324/9781003558439-5

effectiveness), let's now focus on connecting this idea to the concept of framing. Importantly, framing can be useful in all areas of motivating collective climate action – be it through people's social identification, moral beliefs, efficacy beliefs, or other motivations.

Frames are everywhere, in-between the lines of communication, in the unwritten rules of institutions, and even in the form of a collective climate action. Framing concerns the choice of what we say, how we say it, what we emphasize, and what we leave unsaid. Thereby, framing shapes how all of us – often unconsciously – think about, talk about, and perceive what is happening around and inside of us.

The exact definition of framing varies across the fields of economics, politics, social sciences, and ecology. The definition used throughout this book comes from the US-based activism communications organization the *Center for Story-based Strategy*. For them, framing is the design of a narrative with "characters, conflicts, images, and foreshadowing that reinforces a good story and creates meaning for an audience".[1] This definition emphasizes that the process of framing is aimed at reaching people's hearts and minds, and that if we want to understand framing completely it makes sense to look at what frames are made of: stories.

In 1944, psychologists Fritz Heider and Mary-Ann Simmel from Smith College ran a simple but significant experiment to show how human brains are wired to interpret and create meaning through stories. In their experiment, they presented participants with an animation of a pair of triangles and a circle moving around a square.[2,3] They then asked the participants to describe what was happening. Some of the participants described the circle as "chasing" the triangle – thus creating a story out of nothing more than a need to interpret the animation presented to them. This example illustrates how our human brains often inevitably search for and create stories.

Such storytelling can also take place on the collective level. Through history and institutions, groups can create stories that shape cultures and people's ways of establishing collective meaning. As a consequence, it is important to challenge stories that don't serve socio-ecological change, especially if they are silent, harmful assumptions pervasive in the dominant culture.

Similar to what you have now read about collective action, the influence of stories lies in our perception of them and the meaning we give them – much more so than in how things actually happened.

"The way in which the world is imagined determines at any particular moment what people will do."[4]

As researchers, this Author Team strongly believes in the power of knowledge and facts. However, as psychologists, we also see the need to bridge the gap between information and action. In that vein, it may be useful for climate action groups to intentionally create stories that have the potential to influence people's perceptions and actions. That's where framing comes in.

CHALLENGING EXISTING FRAMES

In some cases, it may be sufficient to change how we talk about an issue in order to create a new frame that facilitates a certain response from a target group. For example, research found that US Americans show greater support for reducing economic inequality when it is framed as a *lower-class disadvantage* (and not as an *upper-class advantage*).[5] In other cases, much more than a simple change in word choice may be needed.

Climate action groups can try to analyze and change pre-existing frames. An illustrative example of places where we can challenge existing frames is in how we look at maps.[1] From an early age, we're shown maps of the world. From map to map, the world's continents do not change in size – Europe is always small, North America is always large – and we begin to take these maps as fact. And facts are always neutral, right?

In reality, maps can be anything but neutral. For the two-dimensional maps shown to us to educate us in world geography, both Europe and North America actually appear far larger than they truly are (Mercator projection). Europe is typically placed in the middle of these geographical maps. This disproportion and focus is a Eurocentric framing that can establish perceptions of regional significance, or lack thereof.

Climate action groups can try to alter existing frames. For example, Image 4.1 shows climate protesters holding a banner that reads "There are no jobs on a dead planet". Historically, the climate movement and the labor movement have often clashed over the need to keep fossil fuels in the ground

Image 4.1: Activists carrying a banner with a framed message, Germany (2018).

Photo by Ende Gelände/ Christian Willner (CC BY-SA 2.0)

versus protecting industrial jobs. Today, there is still a common perception of a divide between jobs and the environment. The banner tries to resolve this divide by framing the environment as the basis for providing well-paid industrial jobs.

These examples illustrate the necessity of analyzing the pre-existing frames that shape our view of the world in order to strategically create new ones. Most of the time though, climate action groups run with existing stories and frames rather than creating fundamentally new ones. The next section describes ideas for new frames that could increase the outreach of collective climate action.

HOW WE CAN CREATE FRAMES FOR COLLECTIVE CLIMATE ACTION

Generally speaking, frames are more persuasive when they are internally consistent, supported by reality, and shared by sources that are perceived as credible and trustworthy.[6,7] Three types of frames are typically found in social movements: identifying problems and attributions (diagnostic frames), predicting the likely course of events (prognostic frames), and promoting the desire to achieve something (motivational frames).[6]

A climate action group might apply a diagnostic frame by highlighting the injustices and moral violations of big oil companies. The group might frame the actions of the oil companies by saying, "these companies are fueling an increase of climate disasters, which hit the Global South and future generations the hardest, all while making a profit". Some group names even make use of specific frames. For instance, the climate action group *Just Stop Oil* uses a diagnostic frame by highlighting the problem of oil extraction. Its German counterpart, *Letzte Generation* [Last Generation], includes a prognostic frame in their name by underlining that they are the last generation capable of preventing the climate crisis from escalating irreparably. And the name of the climate action group *Fridays for Future* can be seen as a motivational frame as it emphasizes the group's desire to achieve a future goal. Indeed, in a study on 15 homeless social movement organizations, researchers found that movement effectiveness can be increased by employing diagnostic and prognostic frames.[8]

In addition to these social movement frames, there are frames that are typically recommended for communicating climate change. These may also be transferred to the framing of collective climate action. The book *The Psychology of Climate Change Communication*[9] offers some guidance here.

One of the framing methods recommended in this book is to address multiple motivational types at once. We all approach our goals in different ways – while some of us may strive for an ideal and seek to promote change (focus on promotion), others may be more concerned with what should be done to prevent negative outcomes (focus on prevention). Importantly, both these ways of approaching goals are useful in fighting the climate crisis, though the motivations differ. If collective climate action seeks to motivate a wide range of people, it is therefore useful to frame an issue in terms of promotion (using words such as ideal, hope, desire, progress, and opportunity) *and* prevention (using words such as ought, maintain, responsibility, necessity, protect). Using both frame types could also be helpful when trying to find new and diverse members for a climate action group.

Another framing method is to localize and globalize the climate crisis at the same time by emphasizing the consequences of climate change that connect to our own individual lives and to the bigger picture.[10] For climate action groups, this could mean framing actions in terms of not only global but also local environmental problems and solutions. For example, a global climate march could highlight the steps that a local government needs to take to align with the global Paris Agreement.

Yet another framing method recommended in this book is to focus on present and future losses – a message could read "losing less now instead of losing more in the future".[9] Rather than focusing only on the future, frames should aim to include present losses. Moreover, these loss frames may also be more effective than emphasizing what is to be gained. When considering which injustices to highlight in a protest, a climate action group could, for example, pick out an already existing climate impact and give a voice to the people affected by it. For example, the annual *Wir haben es satt* [We're fed up] protests in Berlin are organized by a coalition of farmers and climate action groups demanding a change to a more environmentally and socially just agricultural policy in the face of current and future climate threats and farm extinctions.

The last recommended framing method we're going to touch on here is linking the climate issue to other issues that are relevant to society. The climate crisis can be framed as an environmental threat, but also as an economic, health, and national security threat. Such framing choices may be particularly important when climate action groups seek to build coalitions with other groups and movements. As previously stated, it remains important that the framing in terms of other threats is supported by reality and that people can easily make the link between these threats.

 Box 4.1: The bottom line

Research shows that social movements can be effective when they use diagnostic frames (what is wrong), prognostic frames (what will happen), and motivational frames (what can we achieve). Moreover, climate issues are more likely to be understood by a target group when they are framed as local, present threats that are associated with current losses as well as future ones, not only to the environment but also to other societally relevant areas, such as the economy, health, and national security. In applying these frames, climate action groups can appeal to both promotion-oriented and prevention-oriented individuals.

Box 4.2: Food for thought – The climate justice frame

If you're interested in a more in-depth exploration of how climate justice can be framed, take a look at the 12-month *Framing Climate Justice* research project that brought together climate justice organizers in the UK.[11] For the authors of this research project, placing justice at the heart of climate change communication is essential. To this end, they suggest highlighting solidarity with those disproportionately impacted while also acknowledging injustices within the UK. Frames should include affected groups as central actors in decision-making processes in a "self-direction is key" vein. The authors also stress the importance of being cautious when applying emergency frames. Many people are already well aware of the urgency of the climate crisis, and overly alarmist messaging might hinder the potential for solidarity. According to the researchers, it may be better, then, to focus on drawing clear connections between climate issues, capitalism, and colonialism.

CONSIDERING TARGET AUDIENCES IN FRAMING DECISIONS

As nice as it would be, there's no one-size-fits-all method of framing. Whether a climate action group is addressing its members, like-minded groups, or people from outside the movement makes a big difference in how a message should be framed in order to resonate with the target audience. In the quest for useful frames, it is therefore crucial to consider relevant characteristics of a target audience and to understand its values.

Framing an issue in terms of people's existing values can make the issue more appealing and impactful. This method, called value-based communication, was discussed in Chapter 2 as a way of linking existing and new groups to the fight against the climate crisis, and Chapter 3 provided a basis for considering which values may be relevant to a specific target group.

A study by researchers from the *Centre for Research on Environmental Decisions* at Columbia University illustrates this point.[9] These researchers surveyed a national sample about a policy program that would raise the cost of certain climate-damaging products and use the money to fund renewable energy and carbon capture projects. The researchers employed two different frames and then measured the support for the program for each framing method. They found that there was significantly more support when the program was described as a "carbon offset" (52%) compared to a "carbon tax" (39%). Perhaps unsurprisingly, a person's political affiliation corresponded to how they viewed these frames. Participants with more liberal views tended not to discriminate between the two frames at all, whereas conservative-leaning participants showed a strong preference for the "carbon offset" frame.[9] Issue framing may therefore be particularly important if a climate action group targets individuals who place less importance on environmental and climate justice values (typically conservative voters).

This study shows how powerful frames can be when they target people's values. The same reasoning also applies to other characteristics of a target audience, such as their socio-demographic backgrounds and their psychological and motivational states. Therefore, a target audience member's social identification, perceived social norms, moral beliefs, emotions, and perceived efficacy are useful to consider when deciding on a framing strategy.

 Box 4.3: The bottom line

For frames to be effective, they must be tailored to the values, demographics, and other psychological characteristics of a target group. As a whole, framing is a method that can help climate action groups make use of the manifold psychological processes addressed in this book.

 Box 4.4: Note – Framing in this book

Framing, like storytelling, can serve purposes both good and bad. It therefore feels important to mention that this book, the stories we tell, and the messages we imply, comprise a contemporary product of our time. Political discourses are constantly being negotiated, and so too are the framings and labels we use and the meanings we draw from them. The projects, people, and places we reference throughout this book bear futures we cannot predict, making it possible that the terms we use today will be outdated or even considered offensive tomorrow.

As you reflect on this chapter, on the power and privilege of shaping our own stories, we ask you to take agency in your interpretation of our words: find your own examples, design your own labels, cross out what feels wrong, and write in what feels right. From its inception, this book was a collaboration among the authors. And we hope you see yourself as a fellow collaborator of this book as well.

References

1. Reinsborough, P. & Canning, D. *Re:Imagining Change: How to use Story-Based Strategy to Win Campaigns, Build Movements, and Change the World.* (PM Press, 2017).
2. Heider, F. & Simmel, M. An experimental study of apparent behavior. *Am. J. Psychol.* 57, 243–259 (1944). https://doi.org/10.2307/1416950
3. Lück, H. E. Die Heider-Simmel-Studie (1944) in neueren Replikationen. *Gr. Organ.* 37, 185–196 (2006). https://doi.org/10.1007/s11612-006-0021-0
4. Lippmann, W. *Public Opinion.* (Harcourt, Brace & Co., 1922).

5. Dietze, P. & Craig, M. A. Framing economic inequality and policy as group disadvantages (versus group advantages) spurs support for action. *Nat. Hum. Behav.* 5, 349–360 (2020). https://doi.org/10.1038/s41562-020-00988-4

6. Benford, R. D. & Snow, D. A. Framing processes and social movements: An overview and assessment. *Annu. Rev. Sociol.* 26, 611–639 (2000). https://doi.org/10.1146/annurev.soc.26.1.611

7. Gulliver, R., Wibisono, S., Fielding, K. S. & Louis, W. R. *The Psychology of Effective Activism*. (Cambridge University Press, 2021). https://doi.org/10.1017/9781108975476

8. Cress, D. M. & Snow, D. A. The outcomes of homeless mobilization: The influence of organization, disruption, political mediation, and framing. *Am. J. Sociol.* 105, 1063–1104 (2000). https://doi.org/10.1086/210399

9. Shome, D. & Marx, S. M. The psychology of climate change communication: A guide for scientists, journalists, educators, political aides, and the interested public. (2009). https://doi.org/10.7916/d8-byzb-0s23

10. Armstrong, A. K., Krasny, M. E. & Schuldt, J. P. *Communicating Climate Change: A Guide for Educators*. (Cornell University Press, 2018).

11. PIRC, 350.org & NEON. Framing climate justice. *Framing Climate Justice* https://framingclimatejustice.org/

5 EFFICACY **BELIEFS**

DEFINING EFFICACY BELIEFS

Efficacy beliefs are a potential source of motivation or demotivation that everyone involved in the climate movement has most likely experienced at one point or another. At times, it can feel like change is truly possible, like climate action groups and their members are really contributing to large-scale transformation. At other times, it can feel like nothing is moving forward, like we're better off hiding away

DOI: 10.4324/9781003558439-6

from the world and just giving up. These are the extremes of the feeling and perception of efficacy, the endpoints on the spectrum of our efficacy beliefs.

The following excerpt is taken from an unpublished interview study conducted with environmentally engaged people regarding their moments of efficacy. In it, a person involved in a seed exchange describes enthusiastically how, for them, positive change can spread and grow just like a seed does.

> It was a seed exchange and there are people who just have a huge box full of seeds [...] and now I have the seeds myself. So now I can multiply the seeds [...], let's say, I can make 40 seeds out of it, then I have a multiplication of 40. So, and that is what is exciting. And when everyone starts multiplying, then nature is something like a catalyst that leads to the fact that you don't feel weak, but that you can achieve something together.
>
> [translated from the original German]

It takes courage to both face your fears of ineffectiveness and to let yourself feel effective in the face of a threat as great as the climate crisis. As described in Chapter 2, a sense of efficacy can be considered a psychological need that, when met, can push us towards action. If our need for efficacy is satisfied, it can lead to a happy and healthy life and can keep us attached to groups that make us feel efficacious.[1] Efficacy beliefs can also be seen as pulling us towards specific goals and keeping up our motivation to pursue them. Accordingly, they motivate our actions in a wide variety of contexts.[2] It is therefore not surprising that efficacy beliefs are also highly relevant to collective climate action.[3-5] Throughout this book, and similar to moral beliefs, efficacy beliefs comprise both perceptions and feelings. Previous research finds that two types of efficacy beliefs are key for involvement in protest and volunteering: collective efficacy and participative efficacy.[6,7]

Collective efficacy

Collective efficacy is the belief that we as a group can achieve our goals through joint action. An example of this is a *Fridays for Future* protester believing that their group can really influence political decision-making. When protesters chant "Power to the people, 'cause the people got the power", this highlights their feelings of collective efficacy. There are as many collective efficacy beliefs as there are groups that we ourselves feel part of.

Participative efficacy

Participative efficacy is the belief that one's own contribution is important for a group and that without that individual participation the group might not achieve its goals. An example of this can be seen in an individual believing that their contribution to their neighbourhood's new energy cooperative is necessary because of their background knowledge, without which the group might not succeed. Participative efficacy is also reflected in the title of Greta Thunberg's book *No One is Too Small to Make a Difference*.[8]

FROM EFFICACY BELIEFS TO COLLECTIVE CLIMATE ACTION

Collective efficacy can be connected to a variety of climate actions, such as joining an environmental protest (like the one shown in Image 5.1), marching in a climate strike, riding one's bike in a protest for a bike-friendly city, signing a petition against coal mining, signing a *Greenpeace* petition, supporting protests in the German Hambach Forest, and participating in a local neighborhood initiative for climate protection, just to name a few.[9-17] Each of these examples of the link between collective efficacy and climate action have been demonstrated through studies, and these studies show how people are more motivated to join climate protests and initiatives if they believe that their joint actions as a climate action group are indeed effective. One study on anti-whaling activism, for example, showed how collective efficacy was relevant for supporting not only normative climate action, such as signing petitions, but also non-normative forms of climate action, such as damaging whaling vessels.[18]

Interestingly, there are also many studies showing that collective efficacy is not that relevant for climate action when participative efficacy is studied simultaneously.[19-23] In these studies, which focused on involvement in *Transition Town* initiatives, volunteer work in the fight against water privatization, camps for climate action, and participation in socio-ecological initiatives, participative efficacy was found to be more vital than collective efficacy.[24-29] While collective efficacy is an important driver of climate action in general, current research suggests that participative efficacy is even more important.

There are two additional ways of looking at efficacy beliefs: building initial efficacy beliefs and maintaining resilient efficacy beliefs.

Image 5.1: Environmental Protest in Detroit, USA (2019).

Photo by Paul Becker, Becker1999 (CC BY 2.0 Deed)

Building efficacy beliefs

Building initial efficacy beliefs might be most important for getting people to join climate groups. (Potential) newcomers really have to feel that they can make a significant contribution to the climate movement and that the actions of the movement will be effective in promoting change. Research shows that it is easier to build efficacy beliefs earlier in life as optimism tends to fade with age.[30] Indeed, so-called youthful optimism helps with overcoming doubts and taking action even in the face of immense challenges. It should therefore be considered important to avoid criticizing the optimism of any newcomers to a climate action group, as it just may have been their optimism that led them to get involved in the first place.

Maintaining efficacy beliefs

All of us who become involved in the climate movement face setbacks and failure at one point or another. An interview study found that the longer a person was involved in collective action, the more effort it took to maintain their efficacy beliefs.[30] Most who are members of climate action groups struggle with efficacy beliefs over time, but it may be comforting to hear that this is a common phenomenon and that they are not alone with this. The findings of the interview study also show how important it is that our efficacy beliefs are flexible and resilient, so they can withstand defeat. Figure 5.1 gives a theoretical example of how efficacy beliefs ideally become more resilient over the course of climate movement engagement.

 Box 5.1: The privilege of feeling effective

Before jumping into strategies for fostering efficacy beliefs, it is relevant to mention the inequalities inherent in the possibility of feeling effective that make efficacy beliefs something that can be regarded as a privilege.

Previous research acknowledges that people can deal with failure more easily if they have resources – be that time, money, social contacts, or support.[31] One study found that those who have a higher socio-economic status and are male are more likely to have stronger efficacy beliefs.[32] Moreover, a just system with less corruption can make it easier to develop efficacy beliefs.[18] And, people in countries with a higher GDP and in which women are more strongly politically represented report a stronger general perception of control.[33]

Where we're born, our socio-economic status, our gender, and even our country's GDP are all closely linked to the possibility we feel we have within us to be effective. So, fighting these injustices becomes an essential part of building and maintaining the efficacy beliefs needed to work towards climate justice.

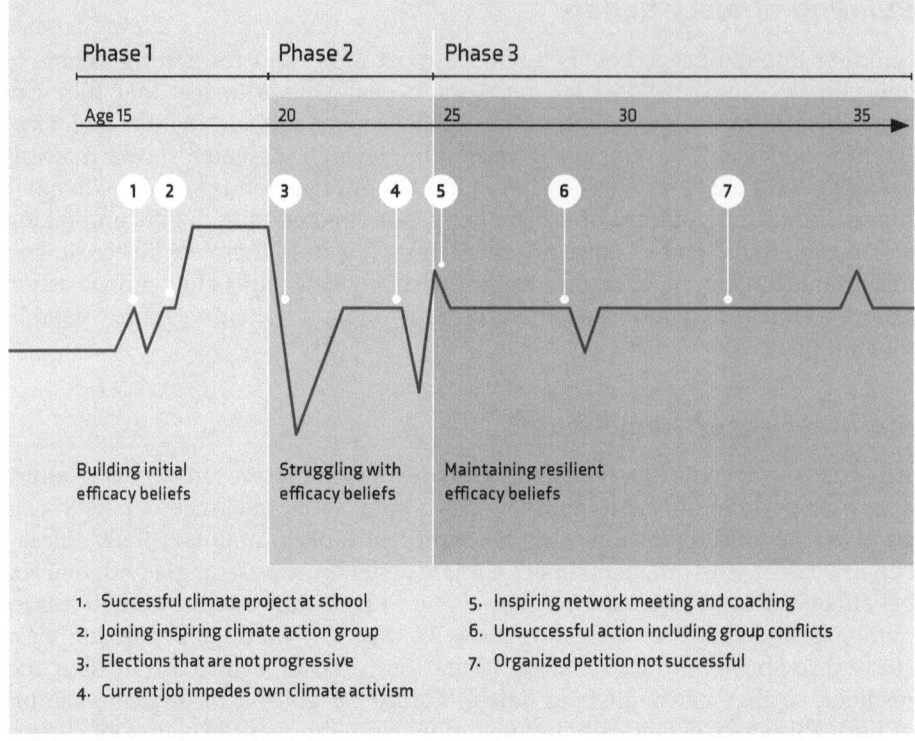

Phase 1   Phase 2   Phase 3

Age 15   20   25   30   35

Building initial
efficacy beliefs

Struggling with
efficacy beliefs

Maintaining resilient
efficacy beliefs

1. Successful climate project at school
2. Joining inspiring climate action group
3. Elections that are not progressive
4. Current job impedes own climate activism

5. Inspiring network meeting and coaching
6. Unsuccessful action including group conflicts
7. Organized petition not successful

Figure 5.1: Theoretical development of a person's efficacy beliefs with higher and lower levels of efficacy beliefs, as well as initial struggles and ultimate resilience

HOW WE CAN FOSTER EFFICACY BELIEFS

Efficacy beliefs should not be confused with actual efficacy or success. Of course, in most cases, we build our beliefs about efficacy on previous successes. However, especially in the case of the climate crisis, success is oftentimes not visible and depends strongly on our own interpretation of a given situation. Just how much "success" depends on interpretation can be seen in interviews with road-blockading activists who developed efficacy beliefs despite being unsuccessful in their core mission.[34] Their failed collective action left them with the impression that action is even more urgently needed and therefore legitimate. Additionally, they perceived other aims as accomplished, such as contributing to building the movement. These perceptions strengthened their sense of efficacy and motivated future actions.[31]

Since efficacy beliefs surrounding climate action do not only depend on success but also on our own perceptions, it is crucial to know how to foster such perceptions. Previous research has shown, however, that it is quite difficult to foster climate- and environment-related efficacy beliefs.[12,35,36] This finding makes a lot of sense: climate action groups often struggle to convince large segments of the population of their joint efficacy, so why should this task be any easier for researchers? Fostering efficacy beliefs is complex, especially as people seem to diverge in the factors influencing their beliefs.

In interviews that are yet to be published, our Author Team found that people engaged in socio-ecological change strongly differ in what forms the basis of their efficacy beliefs. One person drew a feeling of efficacy from the memory of a very successful protest – even though, or perhaps because, the four before it were unsuccessful. Another person maintained their efficacy belief by recycling in the workplace and therefore setting an example; even though they were also involved in crucial political decision-making, this did not give them the same feeling of effectiveness.

Those of us who have long been involved in the climate movement probably already have a few strategies up our sleeves for coping with failure and the coinciding sense of inefficacy. Still, there may be times when it is difficult to use these tactics. With that in mind, and considering that people new to the movement probably haven't needed to develop these kinds of tactics yet, exploring strategies for fostering efficacy beliefs is worthwhile for us all.

In the following, you can get to know several strategies for promoting efficacy beliefs. Focus 1 is about strategies accentuating positive changes. Focus 2 features strategies for nurturing efficacy within a group.

Focus 1: Accentuating positive changes

It is easier for us to get involved in collective climate action if we have perceived positive changes in the past. For example, successful protests can lead us to anticipate similar successes in the future.

 Box 5.2: Sense of efficacy

Here is space to reflect on your own sense of efficacy on your climate journey so far.

Ask yourself:

- *What have been my highs?*
- *In which instances have I felt especially effective in facing the climate crisis?*
- *Was I on my own or with a group?*

Then ask yourself:

- *What have been my lows?*
- *In which instances have I felt like I was losing faith?*
- *How did I regain a sense of efficacy?*
- *How do I currently feel about my efficacy, and what influences it these days?*

You may find it helpful to organize these moments into a trajectory, as in Figure 5.1, which will help to show you where you have come from and where you may be heading.

Focus 1 – Strategy 1: Highlighting success and efficacy

Success can clearly help build a sense of efficacy. This is why focusing on an action's success, its effectiveness, and your own contribution to it is a very important strategy for fostering efficacy beliefs.

Focusing on success

One study from the universities of Leipzig and Queensland had participants all under the age of 30 read a text on the transition to sustainable mobility transition in Germany.[37] Participants were divided into two groups. The first group was assigned a text that claimed individuals under 30 were working together to promote environmentally-friendly mobility behavior, and that their efforts were positively impacting the use of hybrid and electric vehicles in Germany. The second group was assigned the same text, but it stated that the under 30s had not created any change. Participants who read the first text were more convinced that they could, along with others, contribute to a sustainable mobility transition, which in turn was related to engagement in climate action. In this study, even a short text was enough to elicit these perceptions of efficacy.

Efficacy messaging such as this is used a lot by climate action groups. The line "we are the investment risk", used by the climate action group *Ende Gelände* in their video of a coal mine blockade,[38] implies that they as a group can disturb the regular procedures of coal mining. An additional study found that in order to foster efficacy beliefs in others it is useful to focus communication on the chance of success, however small, rather than on the large chance of failure.[39]

Our own engagement with social media can also be a tool for influencing our efficacy beliefs. In many cases, we can influence how often we confront ourselves with stories of success and stories of failure. People involved in collective climate action could therefore ask themselves what type of stories they want to read and how much information on catastrophes and failure they can endure without their perceived efficacy, and consequently their motivation for action, diminishing. While it is not often talked about, it is not uncommon for members of climate action groups to take breaks from reading the news in order to recuperate and rebuild efficacy beliefs. This is why the emerging trend of constructive journalism, highlighting positive news, is very important. One example of constructive journalism is the online magazine *Perspective Daily*,[40] which covers topics such as sustainability, health, education, and society, and explicitly focuses on solutions instead of problems, so that readers gain a better idea of how to effect change in the world.

Not only journalists but every one of us can consider the effect positive and negative messaging has on others and their efficacy perceptions. For example, climate justice activist Tadzio Müller writes,

> As an exponent of the disobedient, anti-growth wing of the climate movement, I have to explain 'our' strategy internally and represent it externally. That means every day that I communicate politically I have to bridge the distance in my mind between my knowledge of the near 100 percent

probability of climate collapse and my hopeful speeches. My means to do this: magical movement realism.[41]

[translated from the original German]

Conveying hopeful messages can take a lot of effort and might also make it difficult to get into authentic contact with others. Thus, it is necessary to carefully weigh the reasons for bringing good and bad news to others.

Linking success to our actions

A number of interview studies suggest that success has even more potential to foster efficacy beliefs if it occurs unexpectedly.[30] One example of this can be seen with the 2010 anti-nuclear protests in Gorleben, Germany. Shortly after the protests, the terrible Fukushima catastrophe occurred, prompting the German government led by the conservative party to rapidly pass a nuclear phase out resolution. Through speaking with protesters, *Wandelwerk* learned that the unforeseen success of a coal phase out positively impacted participants' efficacy beliefs in the long term, especially for those whom this protest was their first activist experience. It is not clear to what extent the normative demonstrations and non-normative blockades, taken together with the Fukushima catastrophe, influenced the government's decision. However, by actively attributing a success to the contributions of a group and its members, groups can build a sense of efficacy.[30,42]

Linking success to our own contribution, however indirect, has its place in fostering efficacy beliefs. As discussed in Chapter 3, Focus 2 – Strategy 4 on considering the impacts of a radical flank, the influence of climate action groups within a movement can be manifold and indirect. Impacts are often hard to detect, making it difficult to feel directly effective. But the invisibility of impacts can also provide flexibility in just how we interpret efficacy.

During the rise of the *Fridays for Future* movement, for example, many newspaper articles questioned why the climate movement had not managed to address and mobilize that many people earlier – what had they been doing wrong?[43] As we on the Author Team experienced, this led to a lot of long-term members of the movement feeling frustrated and as though their previous efforts were being belittled, which in turn negatively affected their efficacy beliefs.

Here's where using an attribution strategy comes in – rather than diminishing the value of previous actions, long-term members of climate groups would do well to look at how their own and their group's contributions led to the rise of *Fridays for Future*. For example, if they were involved in environmental education, it's possible they contributed to motivating students to initiate or join this movement. One could also take the perspective that the pre-existing climate movement had already set the stage and mobilized a society-wide consensus upon which *Fridays for Future* was able to build, before ultimately bringing the masses to the streets. In Germany, for example, large protests took place in Hambach Forest a couple of months before the emergence of *Fridays for Future*.[44]

Within climate action groups, a culture that makes everyone feel that their own contribution is valuable and seen by others is crucial to promote participative efficacy.[28] The following interview statement in the study on the *Slow Food* initiative reflects this: "to get that sort of appreciation back, it was heartening and rejuvenating, like we're in this together".[45] As actual, physical feedback about the impacts of collective climate action is oftentimes invisible, an appreciative social feedback culture is key for building and maintaining efficacy beliefs within groups.[28,46]

> ## Efforts should be acknowledged and celebrated, no matter how successful they might or might not be.

For example, social feedback was provided at the end of a training seminar for sustainability coaches for the climate action group *network n*. Participants were tasked with attaching a piece of paper to each other's backs and then invited to go round the room and write on this piece of paper the potential and skills they saw in each person. Being reminded of their skills in such a way likely boosted their efficacy beliefs and strengthened the bond of the group.

Yet, we needn't always look to others for appreciative social feedback – we always have the opportunity to engage in self-affirmation and to appreciate our own work.[47] Appreciative self-feedback can help to make us less dependent on feedback from others and therefore more resilient in times of feedback scarcity or group conflict.

 Box 5.3: The bottom line

If a climate action group wants to foster beliefs and feelings of efficacy, they need to emphasize already existing successes and explicitly highlight the efficacy of the group's role in those successes. Success is even more motivating if it comes unexpectedly. In order to maintain efficacy beliefs in the face of both success and failure, it is key to develop strategies that attribute success to one's own (group) efforts and build appreciative feedback cultures.

Focus 1 – Strategy 2: Emphasizing that many others are involved

Looking back at 2019, it seems remarkable that the climate action group *Fridays for Future* was able to sustain weekly school strikes for the climate for over a year. Regardless of how you might have viewed this climate action, the strike was an impressive display of collective fortitude and determined action, reinforced even further by its public visibility. The strike frequently occupied the spotlight in news media around the world. All the while, the *Fridays for Future* logo became a common sight, popping up in social media profile pictures and as stickers on traffic signs, lamp posts, and city walls.

Making visible the fact that social norms are shifting towards climate action and that many people are already engaged in climate action is a relevant lever for promoting efficacy beliefs. If we feel that others are inactive, our feeling of efficacy dwindles. If we feel that many others are also standing up, this can be a strong motivator, especially if this feeling is based on a group's unity.[48] Numerous studies show that large numbers of participants in protests or petitions foster efficacy beliefs.[14,15,49,50] It thus seems important to mention the numbers *and* generate the feeling that they are large (descriptive norm) and increasing (norm trend, see Focus 1 – Strategy 3 in Chapter 2 for more on group norms).

In one study, students at the University of Groningen read texts they were told were written by the *Intergovernmental Panel on Climate Change*.[16] All participants first received information on climate change. One group of participants was then told that people across the globe are collectively fighting climate change, while another group was not given this information. The participants who received this additional information perceived more efficacy, which in turn promoted a variety of climate actions. Another investigation on anti-coal mining activism found that protest imagery on climate action can indeed increase the perception that a large number of people are involved, which in turn fosters efficacy beliefs.[14]

However, climate action groups should not rely on a single picture or video for sparking efficacy beliefs. In one study, participants watched activist videos about the Hambach Forest protests.[12] While they clearly perceived one specific video as portraying efficacy, none of the videos actually influenced their personal belief in their own ability to reduce climate change.

What can we take from this? When it comes to campaign material, rather than showing an image of three people at a protest, the image could show three protesters with many others walking behind them. The social justice movement *Occupy Wallstreet*'s narrative, "we are the 99%", also illustrates putting the focus on a larger group. In workshops on environmental psychology, *Wandelwerk* members often use the trailer of the movie *Tomorrow*[51] to demonstrate how viewing large numbers of people practicing socio-ecological alternatives can increase people's efficacy. When organizing events, meetings, and protests, climate action groups could, every so often, join forces with other climate groups so as to give the feeling that there are many people already involved in the movement. What is more, large coalitions like these may also motivate individuals outside the climate movement. For instance, the occurrence of large climate marches positively influenced belief in the efficacy of acting collectively against climate change among conservatives in the US.[52]

When large numbers are involved in collective climate actions, it's important to consider the context of an action, to highlight incidental injustices, and to not forget to provide action guidance to individuals.

Context matters

A study on anti-coal mining and anti-whaling activism found that focusing on normative rather than non-normative forms of protest was more effective in raising efficacy beliefs.[18] This was especially true when an action took place in a context without corruption. However, if corruption such as bribing to legalize whaling was present, non-normative forms of protest were perceived as more legitimate (see Chapter 3, Focus 2 – Strategy 6 for more on designing a climate action so it is perceived as legitimate, relatable, and effective).

In another study, the type of collective action portrayed in the images of campaign material, whether it was a protest or small activist meeting, made no difference in influencing efficacy beliefs.[14] Taking the findings of these studies together, we suggest that rather than the type of collective action in and of itself, the context around a collective action seems relevant for how it is perceived.

Highlighting incidental injustices

The climate crisis is inevitably connected to structural injustices. However, a meta-analysis of collective action in general (including climate action) found that efficacy beliefs relate to action more strongly in contexts of incidental injustice.[53] This may be because it seems easier to fight incidental injustice, while structural injustice takes more time and effort to address. If a climate action group wants to promote climate action through raising efficacy beliefs, it could thus be useful to search for an incidental injustice as a hook when portraying protests and the large numbers of people involved. For example, if a group wants to call for protests against structural climate injustice in the Global South, it might be beneficial to look for an incidental injustice to focus on first – for example, perhaps there is a country in the Global South in which a specific nature protection law has recently been transgressed for the first time (an incidental injustice) and now endangers the local population.

Don't forget to provide individual action guidance

Emphasizing large numbers needs to be combined with providing practical ideas for how individuals can contribute. This is especially relevant for newcomers and outsiders who are not yet involved in collective climate action.

In one interview study, participants were required to watch an episode of *Years of Living Dangerously*, which portrays best practice examples of people already contributing to a socio-ecological transition.[54] Afterwards, one interviewee said, "the information is empowering, but I'm disempowered in that, what can I do about it? I rent an apartment so I can't go out and get solar [panels]." Here we see an example of someone individualizing a structural problem that actually requires collective climate action, the result of which was disempowerment. It is questionable whether a climate documentary alone can provide individual action guidance suitable for everyone. Yet, climate action groups could create contexts where people could watch such material together and discuss where and how they can become active in their local surroundings.

This participant's statement demonstrates how someone can simultaneously feel an increase in collective efficacy while not developing participative efficacy (the belief that they as an individual can significantly contribute to a group). This conflict is connected to a paradox surrounding collective and participative efficacy.[55] If a movement is shown to involve large numbers of people, collective efficacy may increase at the expense of participative efficacy. In other words, some may think their individual contribution is no longer needed. Mobilizing messages should therefore be comprised of a balance of fostering efficacy beliefs regarding the group and fostering efficacy beliefs regarding each individual's contribution to that group.

 Box 5.4: The bottom line

By portraying that many people are already involved in the climate movement and its groups, efficacy beliefs can be positively influenced. Focusing on incidental injustice and providing guidance on how to become personally involved seem central.

Focus 1 – Strategy 3: Eliciting positive feelings and hope

You might be familiar with that exciting rush of feeling like you can achieve anything – in your personal life, in your job, or even with regard to socio-ecological change. As much as efficacy beliefs are connected to thoughts and perceptions, they are also connected to feelings and emotions.

> Scholars have proposed that, in order to promote efficacy beliefs, people need to experience emotional reactions rather than be confronted with rational reasons.[56]

Emotional reactions can be feeling enthusiastic, strong, proud, or moved, and they in turn promote climate action.[12,24,28,42,57,58] Notably, fear does not seem to affect people's belief in their own efficacy.[16,59] However, those experiencing fear in the face of climate change may be encouraged to act when that fear is coupled with the idea that they have the means to face it (with efficacy beliefs).[60]

One of the emotions most strongly connected to efficacy beliefs is the feeling of hope.[24,28,61] Previous surveys indicate that of those who accept the reality of climate change, more than 50% feel the situation is hopeless.[54] But, hope comes in many colors – in the context of climate change, many of us may hope that the consequences of climate change won't be as severe as predicted. In its extreme form, however, this hope can overlap with forms of climate change denial[62]. We may hope that technical solutions are invented, but if we rely too much on technical solutions, this has the potential to lead to inaction.[63]

Or we may hope that socio-ecological change is possible. This kind of hope, sometimes referred to as "constructive hope",[64] is inextricably linked to efficacy beliefs. One of our authors still remembers the feeling of watching a video of an *Ende Gelände* action she had participated in. For her, the video sparked feelings of both deep desperation regarding the current state of the world and, at the same time, deep appreciation for everyone who was fighting for a just and sustainable world. Despite the resistance to change being so large, the video ultimately left her with a feeling of being moved by the possibility of change (hopefulness) she saw therein. It gave her a sense that this group and its actions could have a meaningful impact.

Findings are mixed on whether hopeful messages about the climate crisis actually lead to perceptions of efficacy or collective climate action.[23,65-67] One study found that efficacy beliefs only influenced action when people were also hopeful,[61] which highlights hope as a necessary condition for efficacy beliefs. So, in terms of climate messaging, it may be a good idea to focus on both the idea that change is possible and the idea that individuals and groups can contribute to this change.

Related to hope, efficacy beliefs may also be strengthened by the belief that social systems are flexible and can change.[68] For example, though the Covid-19 pandemic led to major setbacks for the climate movement, it also showed that governments can initiate many and significant societal changes in a short period of time. From these changes, we can glean how flexible society and governments can be when push comes to shove. If we look at climate change through a similar lens, we might find that the political system is much more flexible in making regulations and providing resources than previously thought, which might elevate our belief that we can effect change for the better.[69]

Climate action groups can reflect on what gives them hope and what they are thankful for. And they can look for and communicate solutions rather than get stuck in problem analysis, in order to be effective[70-72]. It may also be helpful to discuss what socio-political institutions are perceived as the most flexible and therefore changeable. Climate action groups can also plan actions that make their members and potential newcomers feel enthusiastic, strong, and moved.

One example of an action that was carried out with the purpose of keeping spirits up (and in turn helped to maintain efficacy beliefs) comes from a Germany-based camp for climate action. On the campsite, a few individuals decided to run a crêpe stand, where they cooked and served the sweet treat to fellow climate action participants. This spread joy and reminded the activists of the value in treating themselves and others well. From the experience of the Author Team, group methods such as energizing tasks, games, and emotion-sharing rounds can also create positive group feelings (see Focus 2 – Strategy 1 in Chapter 2 for more on fostering a sense of belonging and fun).

 Box 5.5: The bottom line

Fostering efficacy beliefs means creating spaces for positive emotions like hope, enthusiasm, and powerfulness to flourish in climate actions and within groups.

Focus 1 – Strategy 4: Envisioning a better socio-ecological future

Imagine society in 2050. Here, people live in renewed, isolated accommodations, sharing with their neighbors specific areas such as gardens, conservatories, garages, laundry and utility rooms. […] Generally, in this 2050 society, people commute less as they live close to their job and their family. Thermic vehicles are replaced by electric vehicles that are socially shared. Trips are often organized via car-sharing. Train travel becomes the most common way of travelling for day-to-day activities. [...] Employment is re-organized in rural local areas and in small and medium sized cities, people living in big metropolitan areas move to the countryside. Job offers increase in domains such as handicraft, agriculture, and home help, and employment in general is organized around the needs and issues of citizens at a small local scale (neighborhood, village, small city). People's profession and professional choices largely depend on the local needs. With regards to citizens' dietary habits, their diet follows the seasons, they prefer local products, frequently cultivated in shared gardens. People eat bio produce, less transformed foods and especially less meat. Consumption is in general based on eco-sufficiency, also thanks to the development and implementation of collaborative and sharing services, second-hand and reparation services. When citizens need an object, a tool, an appliance, they rent or borrow it from friends and neighbors rather than buying it new.

(Excerpt from a study by Lucia Bosone and colleagues
from the universities of Paris and Gustave Eiffel, in which this
vision managed to increase efficacy beliefs)[73]

Research shows that one way of eliciting positive feelings and promoting efficacy beliefs and collective climate action is to orient ourselves towards positive visions of the future.[28,50,73–76] Such visions are a desired version of the future,[77] as in the Bosone and colleagues extract, and may include ideas of what we want society and ourselves to be – our own values in action (see Chapter 3 for more on core values).

> Visions are important because they create motivation for socio-ecological change, set standards against which we can evaluate and criticize reality, and make it possible to escape reality once in a while.[77]

This is illustrated by a quote from an activist at the G8 protests:

the area that the camp occupied became [...] the small little island of sanity amongst our world; you really got to see an example of how society could be organized. So that made the ideals of what you were fighting for somewhat more tangible and therefore more real.[78]

Many long-term members of the climate movement have experienced their own "island of sanity" at some point. Creating these spaces for newcomers and emphasizing a group culture of visions, solidarity, and authenticity could be highly effective in getting people involved in climate action.

Of course, we are not suggesting talking about visions alone is enough – it is also necessary to act towards attaining ideals of democracy, solidarity, equality, and sustainability. Indeed, in an unpublished study, one of our authors has found that camps for climate action elicited visions of better societies to a stronger degree if participants saw and experienced actual solidarity during their time at a camp.[79]

Generating, discussing, and focusing on visions seems to be a recent trend in the climate movement. Envisioning is reflected in the climate protest chant "we are unstoppable; another world is possible". When it comes to what sort of visions should be focused on, studies have found that sustainability visions are typically perceived as more positive and progressive and the imaginary people in them as warmer and more competent than in other types of visions (such as technology visions).[74,80] These traits are what makes sustainability visions particularly motivating for people to contribute to them.[74] When using visionary ideas in speeches or group processes, actions or campaigns, there are some points to consider: visions may be especially effective when they depict particular characteristics of people, use backcasting, and apply methods that fit a target group.

Research shows that visions are especially motivating if the people in these visions are perceived as warm and moral.[80] From the perspective of one of our authors, an example of this can be seen in the prolific sci-fi writer Ursula K. Le Guin's novel, *Voices*. Setting aside that *Voices* is set in a fantasy world, the author perceives Le Guin's characters as positively visionary in how they manage to behave with kindness and appreciation in the face of a terrible war – she viewed them as warm-hearted and therefore inspiring. Next to warmth and morality, it also seems relevant that people in sustainability visions are perceived as competent in order to be considered motivational.[80] What this shows us is that a vision might be considered motivating not just because it involves the structural change of something like a political system but because that change positively influences people's characteristics, needs, and relationships.

Moreover, it seems useful to first create a vision and only in the next steps perform a reality check to work out how to get to this vision – this method is called backcasting.[77] This fits ideas that are presented in future-workshops: it is much easier to brainstorm a vision of the future if current societal problems are not an immediate focus. In fact, given that visions are created by contrasting pre-existing knowledge with possible scenarios, it is likely that anyone trying to envision a future is already thinking about current problems. So, there is no need to bring extra attention to current issues while creating visions.

The process of envisioning may not work for everyone. Recent research has found that visions are especially motivating for people who perceive climate change as a close threat and who see themselves as environmentally-friendly, or, in other words, those who are already convinced.[73] As well, there is initial evidence that some envisioning methods might be overwhelming for some. In a yet unpublished study by one of our authors, participants imagined a journey to their own future vision of a sustainable neighborhood.[81] This "dream journey" motivated climate action in people who had already had preliminary visions of a sustainable future in their minds. For people with no or fewer preliminary visions, the dream journey was instead demotivating, possibly because it asked too much of them. Telling some people to "journey" to their dream society may just lead them nowhere. Instead, these participants were motivated by videos showing how a sustainable future could look, possibly because this sparked ideas. It might therefore be useful to use envisioning methods in a targeted way and allow people who already have a vision to reflect freely on this while showing those with fewer visions what exactly a sustainable world might look like.

 Box 5.6: The bottom line

Envisioning is one method for creating positive emotions and a focus on solutions. Such visions can be most motivating when people are portrayed as being warm, moral, and competent and when a reality check is postponed to after a vision is brainstormed.

Focus 2: Designing group contexts that nurture efficacy

In order to feel more effective, we can of course actively try to look for and create groups that provide this feeling. Agentic groups may help us regain a sense of efficacy if we're feeling helpless.[82] This is why one of the most important elements of effective climate action is going beyond individual action and joining groups that reflect our values, making it possible to achieve much more than we would on our own. But this is easier said than done – a lot of people struggle with finding effective climate action groups or movements they want to belong to. Many may also question the efficacy of a climate group they've already joined. That's where the next strategies on group motivations, sizes, goals, actions, skills, and roles come in.

Focus 2 – Strategy 1: Considering diverse motivations and group sizes

To confront personal struggles with the limited efficacy of groups, it may be crucial to build a group culture and narrative focused not only on efficacy but on diverse motivational pillars. In other words, independent of the effectiveness of a group's actions, group members could focus on, for example, their shared moral beliefs (see Focus 1 in Chapter 3 for more on creating anger-eliciting situations) and positive relationships (see Focus 2 – Strategy 1 in Chapter 2 for more on the need for belonging) with one another as the core features that bind them together. In fact, studies suggest that by fostering social identification and a connection with other group members, a basis for efficacy beliefs can be established.[5,9,53]

In the previously mentioned study which brought participants together to interactively develop strategies for providing clean water to people living in the Global South, participants strengthened their identification and increased their collective efficacy beliefs.[83] Other studies have also found that efficacy is less important to individuals who already have strong moral beliefs.[36,84,85] Thus, the suggestion to build motivation on several psychological pillars is also highly relevant for individuals in addition to groups – each of us who is involved in collective climate action could benefit from asking ourselves:

- *How dependent is my own motivation on success and efficacy?*
- *What else motivates me that could carry me through times of failure?*

The structure of a climate action group also influences people's efficacy beliefs. Large groups may foster a feeling of collective efficacy because the amount of people makes them appear agentic. Smaller groups on the other hand may be more likely to evoke participative efficacy, as an individual's contribution has the potential to be more impactful.[2,86] The question of what the perfect group size is for collective climate action was also raised in the paradox of collective and participative efficacy mentioned in Focus 1 – Strategy 2.

An unpublished study from one of our authors recently investigated how group size affected people's perceptions of energy communities.[87] Energy communities like those depicted in Image 5.2 produce, share, store, and distribute renewable energy. Large energy communities were perceived as more collectively effective, which substantiates research from other fields.[88] Smaller energy communities were considered more coordinated and open to individual contributions. Yet, our study found no overarching preference for supporting either small or large communities.[87]

As far as we, the Author Team, know, there are currently no published studies on the optimal group size for developing efficacy beliefs among members. However, considering the efficacy paradox, it is possible that the optimal group

Image 5.2: Energy initiative at Rainshadow Community Charter High School in Reno, USA celebrating their solar panels (2013).

Photo by Black Rock Solar (CC BY 2.0)

structure for fostering efficacy beliefs is one that is strongly nested – small climate action groups working as part of larger action networks working as part of the movement for socio-ecological transformation.

One example of a nested group is the Author Team's climate action group *Wandelwerk*. The structure of *Wandelwerk* is loosely sociocratic, and our group is comprised of five smaller working groups. At the small-group level, each group autonomously handles their own project (e.g., developing a social media presence) but is advised to discuss larger decisions at "coordination circle" events, which every *Wandelwerk* member can participate in. Of course, like many other organizations, we struggle with non-hierarchical aspirations and the feeling that we are too few to live up to our true potential. However, our small size can boost participative efficacy feelings among our members – every single member is needed and appreciated. At the mid-size-group level, *Wandelwerk* is part of a larger environmental psychology community that consists of numerous organizations and groups, including our parent initiative, the *Initiative Psychologie im Umweltschutz* [Psychology in Environmental Protection Initiative],[89] and *Climate Outreach*[90]. At the large-group level, we at *Wandelwerk* see ourselves as part of the climate justice movement, and we frequently look for opportunities to join forces with other like-minded groups.

Cultivating a nested group structure comprised of groups broadly ranging in size might foster beliefs of participative efficacy and collective efficacy at various levels and at various times. A nested group structure may also make it easier for members to join the collective climate actions of allied groups and broader networks. It could combine the best of both worlds. However, this proposal has yet to be tested.

 Box 5.7: The bottom line

When it comes to fostering efficacy beliefs, a group and its structure matter. Since success in collective climate action is not a given, it might be helpful to build people's motivation on different psychological pillars. Moreover, as large groups seem to increase beliefs of collective efficacy and small groups seem to increase beliefs of participative efficacy, it may be useful to employ nested group structures – small groups acting within mid-size communities acting within large movements.

Focus 2 – Strategy 2: Choosing goals and actions wisely

The goals we explicitly select or implicitly carry in our groups, as well as the actions that we deem effective and therefore undertake, are an essential part of our efficacy beliefs. Therefore, which goals and actions we choose strongly influences our perception of what we can collectively achieve.

Choosing goals

Goals are a concrete reflection of a group's core values and morals. Research suggests that clear and self-determined goals may be a good foundation for efficacy beliefs.[91,92] *Fridays for Future*, for example, managed to establish clear goals within months of its arrival on the public stage. The group formulated clear demands in line with the global temperature increase target set by the Paris Agreement and thereby with what governments had already agreed on. In Germany, for example, the group additionally called for concrete action aimed at closing all coal-fired power plants by 2030 and for the entire energy supply to be switched to renewables by 2035. To the Author Team's knowledge, previous climate action groups had avoided making such concrete demands. The simplicity and plausibility of *Fridays for Future*'s demands made the group's goals very clear. If there were ever any ideological disagreements within the group at this early stage of formation, they were clearly not carried out publicly.

Interestingly, interview studies found that efficacy beliefs are not only reducible to objective success.[34,78,93,94] This may be because our goals determine what we subjectively experience as success. Interviews with members of the anti-capitalist movement at the 2005 G8 protests showed that participants had very diverging definitions of when the protest event could be considered successful, which in turn influenced their efficacy beliefs.[78] While for some, success meant disrupting the summit and blocking the road, others saw success in the simple act of raising awareness.

Climate action can have many goals. If we look at the action of joining a protest, for example, the goals could be to put an end to something, to convince policymakers to do something, to learn as a movement, to build coalitions, or to simply make use of one's right to protest as an end in itself.[95] Some groups have innate connections to certain goals, a few examples of which can be seen in Table 5.1.

Of course, goals are not set in stone. If a group changes their mission statement (who they are), their goals (what they want) might change accordingly.[94] Let's say there's a climate action group that was founded as an initiative to promote a clean neighborhood by organizing trash clean-ups. After a while and some successful events, some group members learn more about the climate

Table 5.1: Examples of groups and their main goals

Group	Goal
People for the Ethical Treatment of Animals (PETA)	animal welfare and an end to animal exploitation
Viva con Agua [Living with water]	clean drinking water that is accessible to all
DivestInvest	a zero-carbon economy through divesting from fossil fuels and investing in climate-friendly solutions
The United Nations	limiting the global temperature rise to 1.5 degrees

crisis and environmental pollution. At the same time, the group attracts new members who would like to focus on systemic change instead of only fighting the symptoms. These newly involved members band together to shift the group's goal from just having a clean neighborhood to preventing waste overall. This change in goals inspires a change in the group's collective actions – they go from local trash clean-ups to petitioning and advocating for a ban of single-use products in their city.

Even without an overall change in group goals, the goals of individuals involved in collective climate action might change. For example, researchers found that, over time, participants in the G8 protests adopted different goals.[78] Things the participants were actually accomplishing, like strengthening the movement, became more focal goals. This was especially the case when they engaged in post-protest discussions with fellow participants.

In the same manner, climate action groups can use goal diversifying strategies and change their goals regarding specific actions and groups in order to maintain a sense of efficacy. Researchers have yet to determine which are the most useful goals for people involved in climate action groups under particular circumstances. Nevertheless, there is reason to believe that having a couple of short-term and long-term goals, as well as back-up goals in case of failure, could be valuable.[96]

Research on private-sphere pro-environmental behavior actually finds that large goals keep people from resting on little achievements and make them look for further strategies to reach their goal.[97,98] However, it may also be helpful to lower expectations of direct success and to reflect on a couple of smaller goals that are important and that might eventually feed into the larger aim.[30] As having an abundance of goals might also demotivate action, for one action cannot possibly achieve all goals at once,[99] it might be best to have just a few but diverse goals.

Establishing a diverse set of goals could be particularly helpful in the early stages of movement engagement. As mentioned before, youthful optimism can be a good lever for efficacy beliefs and climate action. However, the higher we aim, the harder we can fall – but diverse goals could provide a safety net and make it possible to lower expectations if needed. So, when communicating with newcomers about collective climate action, instead of poking holes in their larger goals, as this may kill their motivation, try offering additional and smaller goals that are more achievable.

Essentially, if a person wants to uphold their efficacy beliefs, it could be beneficial to reflect on what goals they want to pursue and how achievable these goals are. Indeed, it seems more realistic to effect change in a smaller context than in a larger one, since in a smaller context, fewer people need to be convinced of a cause. For example, a study by one of the authors found that students at smaller universities reported stronger efficacy beliefs regarding their involvement in sustainability initiatives than those at larger universities did.[28] Given these findings, it may be beneficial for you or your group to reflect on the following questions:

- *In which context are you aiming to effect change?*
- *Is this realistic?*
- *Is it possible to target a smaller context first and upscale your action later on?*

Choosing actions

It might be beneficial to come up with a range of actions that could achieve a certain goal –[99] actions that could be tactically switched as barriers are encountered. It is not uncommon for a group to form around one collective climate action. For example, the main action of *Fridays for Future* is school strikes, *Ende Gelände* formed around the action to occupy coal mines, and *critical masses* revolve around bike-riding protests as a key action. While these forms of collective action might experience initial success, there could come a point at which they fail or do not seem to be enough.

Studies on non-climate related actions show that when people believe that the government is actually responding to people's demands and that the societal system is working, their inclination to join collective action can wind up reduced.[100–102] In turn, losing faith in the efficacy of a group or the government's ability to respond to its citizens may lead people to pursue more non-normative collective action.[101,103] Indeed, a recent study on climate activism found that people's acceptance of non-normative action was higher when they perceived strong climate injustice and a low efficacy of the *Fridays for Future* movement.[58] This may encourage the establishment of radical flanks in climate action groups. Or it could encourage group members to change strategies altogether. It is at this point that low efficacy beliefs may actually be considered useful:

> when the potential for achieving our goals is threatened, low efficacy beliefs are our internal alarm bells signaling that something's got to change.

To identify actions that are within the realm of possibility, it might be helpful for groups to discuss their shared moral foundation and broader aims. This could be part of the process of writing a common vision and mission statement (see Focus 2 – Strategy 3 in Chapter 2 for more on establishing clear meaning and purpose). Based on these, group members could then come up with some other collective climate actions that help them achieve their goals, making the group more flexible when faced with setbacks or stagnation.

 Box 5.8: The bottom line

While the setting of clear goals is crucial for climate action groups, it is important to remember that goals are not set in stone. By diversifying climate goals and climate actions, groups can build a safety net for cases of stagnation or failure.

Focus 2 – Strategy 3: Highlighting and building skills

Our skills and how we perceive them are crucial in building efficacy beliefs. If we perceive ourselves as having skills, we tend to also believe we can achieve something with them. This is also true if we perceive others as having skills – we also tend to believe we can achieve something together.

In a study on student-led sustainability initiatives, group members felt more participative efficacy if they perceived themselves as having the skills needed for the type of collective climate action they were involved in.[28] In this study, project management skills and context-specific knowledge about sustainability at their university were most strongly related to participative efficacy. Similarly, individuals who perceived fellow initiative members as competent reported more collective efficacy (for more on coaching for sustainable transformation, check out Box 5.9).

 Box 5.9: A coaching program for sustainable transformation

In the course of her Ph.D. work, one of our authors studied *Wandercoaching*, a coaching program run by *netzwerk n*.[28] In this coaching program, two trained peer coaches visited sustainability initiatives at higher education institutions for a weekend. Throughout the weekend, groups were guided through team building measures, created a shared vision for the university, discussed their communication structures, were trained in project management, addressed group conflicts, and much more.

Surveys were distributed to and filled out by 317 students from 36 student initiatives four weeks prior to, two weeks after, and six months after the coaching program. Findings from these surveys show that students reported stronger social identification, visions, perceived individual and group skills, efficacy feelings and beliefs, and more collective climate action after the coaching than before it. On average, student groups shared more climate-related events on Facebook during the 1.5 years following the coaching than before it. This was especially true for educational events. This study demonstrates that extensive coaching may help boost a group's spirits so members can identify and cultivate their skills.

If you are interested, the coaching methods used in the coaching program are available online in German.[104]

There are two primary approaches to skills within climate action groups: highlighting them and building them. Highlighting skills is the process of making members more aware of the skills they already have. The *netzwerk n* coaching program, for example, involved a task called "competency figure design". In this task, individuals were asked to draw a stick-figure representing themselves on a piece of paper and write down reflections on what they're good at, what they like to do, what their core values are, and what they want to learn, to share with fellow group members. Participants were then asked to work together to design a group competency figure that combined the most common or central characteristics from their individual figures. Especially in newer groups, this may be a useful tool for members to gain confidence in fellow group members.

Building skills is the process of developing any skills that might be missing in a group. This process is important not only for the sake of skill development but as a means for strengthening efficacy beliefs. For example, a group might want to build transformative leadership skills. In one study from the field of organizational psychology, transformative leaders were characterized by the following features: able to express efficacy to others, focused on team values and team purpose, dedicated to conveying high standards, and engaged in coaching fellow team members.[105] At the group level, members of climate action groups can support each other in their efforts to develop new skills such as these.

At the individual level, tasks of moderate difficulty are most effective in enhancing efficacy beliefs,[106] so skill development should be adapted to individual capabilities. The best results for both individuals and groups will probably occur when skill support comes from a larger institution. For example, the *netzwerk n* coaching study found that students who attended universities with a designated office for environment-related topics reported more efficacy than students at universities without such an office.[28]

 Box 5.10: The bottom line

The competencies that people perceive themselves and others as having are key to promoting efficacy beliefs. Climate action groups would therefore do well to promote group activities that highlight existing skills as well as those that build new ones.

Focus 2 – Strategy 4: Distributing roles

In addition to highlighting and building skills, it can be useful to teach everyone certain skills in order to be able to rotate roles within a climate action group. It might also mean finding specific roles and tasks for individuals which facilitate the development of participative efficacy.

While the Author Team doesn't know of any research on roles in climate action groups, there are a number of interesting heuristic models that can be connected to psychological ideas. For example, there is the four-player typology by the US systems psychologist David Kantor, in which a person can assume one of four different group roles: *mover, supporter, opposer,* or *bystander*.[107,108] While this model is strongly simplistic and is based on weak empirical evidence, it still provides a highly interesting practical basis for reflecting on group roles. With that in mind, let's take a deeper look at these four groups.

Movers are good at initiating action and bringing in a lot of energy, but they're more susceptible to becoming too powerful and controlling.[108] A mover's core motivation may be success and associated efficacy beliefs. Without movers, groups would probably not even start planning actions as there would be no driving force. *Supporters* are vital in progressing ideas and maintaining positive group culture, but they also tend to avoid conflict and can become too compliant.[108] Supporters may be most motivated by their need for identification and relatedness. *Opposers* challenge ideas and provide critical feedback.[108] However, they are rarely satisfied and can end up demotivating others. They may have a strong need for autonomy and be motivated by a desire to voice their moral beliefs. *Bystanders* are like the line to a bigger network, offering new perspectives and silent support.[108] However, it is often unclear whether they are actually part of the group or not. They may seek autonomy and efficacy but don't want to be too strongly identified with a group.

One core message of this typology is that each of these roles has advantages and disadvantages. Also, what we suggested as a role's potentially strongest need could also be seen as their greatest fear; for example, a mover may fear failure and a loss of efficacy, and a supporter may fear being rejected by the group. Discussing these roles can provide clarity when starting out in a project or struggling with a current project. However, as this is just a heuristic model for which the authors have suggested links with psychological concepts, these roles should invite discussion rather than serve as clear-cut scientific categories.

People can also assume different roles at different stages within the climate movement. A heuristic typology by social change activist Bill Moyer distinguishes *rebels*, *reformers*, *change agents*, and *citizens*, all of which are crucial to social movements.[109] According to this commonly used typology, social movements need all four of these roles at different stages of the political process in order to be successful. To that end, awareness of the necessity of all four roles can foster cooperation among different social actors – so, let's take a quick look at them in Table 5.2.

This list is not exhaustive and can be extended to include, for example, journalists who disseminate relevant information[110] and parents who, by educating their children on relevant information, shape the basis for future social change. Other authors have proposed to group individual roles of people and organizations within a movement according to their association with corporations (*mediator, bridge, independent, captive,* and *isolate*),[111] or their systemic function (*acupuncturist, questioner, broker, gardener*)[112].

There are numerous collective climate actions and thus countless opportunities to find the right role for every individual involved. When we think of collective climate action, we tend to picture *rebels* – banner-waving protesters blocking a road. However, movements for a socio-ecological transformation involve diverse roles and often include several unique subgroups, such as bands,

Table 5.2: Moyer's four roles within social movements

Role	Typical action
Rebels	are at the front lines, protesting for change
Reformers	tackle the political system from the inside, while often holding official positions
Change agents	constitute the grassroot movements that promote and educate others on the change they want to see in the world
Citizens	back a movement as voters and silent supporters

cooking teams, and press units. An example is the subgroup *Scientists for Future*, who describe their group role as one that "supports the global climate movement by providing facts and materials based on reliable and accepted scientific data to activists, politicians, decision makers, educators, and the general public"[113]. Subgroups like these allow people to combine their personal passions (e.g., music, culinary arts, writing) or their professions (e.g., scientist) with climate action.

To present people with all these different roles might be overwhelming. Yet helping them to find out what actually suits them could be a real lever of change. Especially when dealing with newcomers in a climate action group, it may be a good idea to ask them how they would like to contribute and what ideas they might have – leaving open the possibility they may just want to act as a supporter at first.

In order to gain a better understanding of which role an individual might prefer within the socio-ecological transformation, more extensive action seems appropriate. An open climate group day, for example, could give individuals the opportunity to visit, observe, and participate in the work of various local initiatives. By having the possibility to look into various things, it may be easier to get a feeling of which roles feel most effective and in line with personal values and ideals. For individuals already part of a group, the previously mentioned "competency figure" makes it possible to learn about everyone's skills and interests and the roles they want to take. In interviews on a sustainability course, half of the students stated that they felt empowered because they could express their own ideas, qualities, and skills.[114] Finding and identifying with roles that fit one's individual needs may even enhance these benefits for people's participative efficacy.

There are certain advantages to distributing tasks not only according to existing skills but also according to learning potential. A study on the environmental organization *Yellow Creek Concerned Citizens* showed that it might be useful not to let group members stick too much to tasks and roles that they already know and feel comfortable with. Here, female group members took up climate action tasks they had not done previously. This increased their efficacy beliefs and made them question traditional gender roles in general.[115,116] Moreover, recent research found that when people are instructed to "just follow their dream" in thinking about their career paths, they tend to go into more gender stereotypical directions.[117] Thus, in addition to asking what skills people want to bring to a climate action group, it seems essential to find out what skills they want to learn. This also means making a diverse range of roles accessible to group members,[118] to aid them in making progress and developing new skills.

 Box 5.11: The bottom line

Most people bring a lot of skills to collective climate actions which may not be visible at first glance. It is therefore important to use methods that highlight everyone's skills and to make it possible to find suitable group roles for everyone, such that everyone feels they can actually make a difference within a climate action group.

DISCOVERING YOUR EFFICACY BELIEFS

In addition to learning all about efficacy beliefs, it may be beneficial to try out and experience some of the mentioned strategies in real life. To that end, take a look at this task designed to help you identify your own competencies and aspirations, both of which contribute to building your efficacy beliefs. This task, adapted from a method developed by *netzwerk n*,[104] can be performed on your own or together in a group with or without the help of a moderator.

Step 1: Getting an overview of your individual competencies

1. Draw yourself on a piece of paper – this can be realistic or an abstract depiction.
2. Write down the answers to the questions shown in Figure 5.2. It may feel awkward to write down so many positive aspects about yourself, but this is an important step as doing so may help you discover aspects about yourself you are not yet aware of. If this step does not come easily to you, try reaching out to people you are close with to see what they think.
3. If you're doing this task in a climate action group, you may also want to add the following question to give you the opportunity to redistribute roles within your group: *What tasks and roles would you like to hand off?*

You have now created your personal competency figure. You may feel inclined to display it somewhere you can see it often or keep it in a drawer to be used as a reminder in times of crisis. Or, you may want to share it with fellow members of your climate action group.

Step 2: Getting an overview of your group's competencies

1. If you drew your competency figures as part of a climate action group exercise, you can now share them with one another. A group size of 3–5 typically works best for maintaining an atmosphere of intimacy. Discuss one question at a time, giving each person the chance to share their own answers. If people are open to it, you could also suggest skills that others have as part of an appreciative group culture, and especially if people have difficulty identifying their own skills.
2. Draw a group competency figure. This should include all aspects of the individual competency figures that are central to the group or have been raised by several group members. Ensure that each group member contributes to this figure.

You have now created your group competency figure, which should give you an impression of what your group's vision is, what collective climate actions reflect your group's skills, what experiences your group members need in order to feel effective, the level of role satisfaction, how to possibly re-distribute roles, and which roles might be missing for achieving your goals. These points can serve as the basis for valuable discussions in your climate action group.

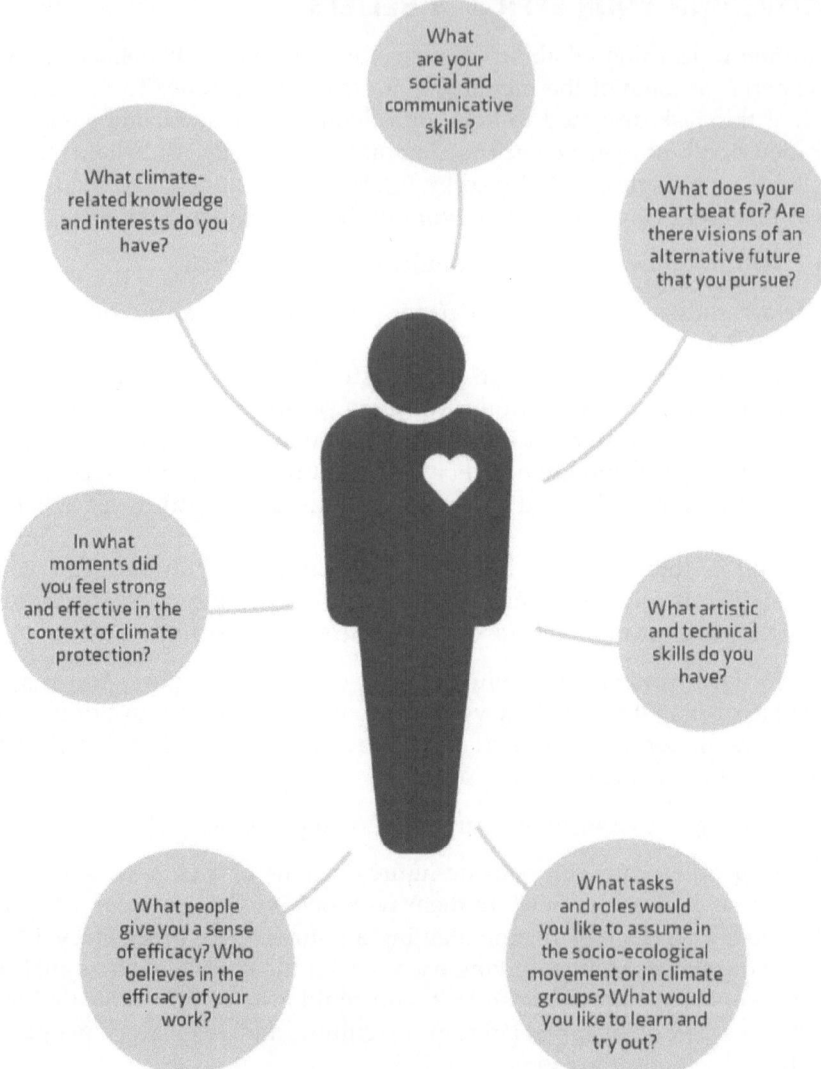

Figure 5.2: Competency figure based on the method developed by *netzwerk n*

References

1. Deci, E. L. & Ryan, R. M. The 'what' and 'why' of goal pursuits: Human needs and the self-determination of behavior. *Psychol. Inq.* 11, 227–268 (2000). https://doi.org/10.1207/S15327965PLI1104_01
2. Bandura, A. *Self-Efficacy: The Exercise of Control.* (W.H. Freeman, 1997).
3. van Zomeren, M., Kutlaca, M. & Turner-Zwinkels, F. Integrating who "we" are with what "we" (will not) stand for: A further extension of the *Social Identity Model of Collective Action. Eur. Rev. Soc. Psychol.* 29, 122–160 (2018). https://doi.org/10.1080/10463283.2018.1479347

4. van Zomeren, M. Building a Tower of Babel? Integrating core motivations and features of social structure into the political psychology of political action. *Polit. Psychol.* 37, 87–114 (2016). https://doi.org/10.1111/pops.12322

5. Agostini, M. & van Zomeren, M. Toward a comprehensive and potentially cross-cultural model of why people engage in collective action: A quantitative research synthesis of four motivations and structural constraints. *Psychol. Bull.* 147, 667–700 (2021). https://doi.org/10.1037/bul0000256

6. van Zomeren, M., Saguy, T. & Schellhaas, F. M. H. Believing in "making a difference" to collective efforts: Participative efficacy beliefs as a unique predictor of collective action. *Group Process. Intergroup Relat.* 16, 618–634 (2013). https://doi.org/10.1177/1368430212467476

7. Hamann, K. R. S., Wullenkord, M. C., Reese, G. & van Zomeren, M. Believing that we can change our world for the better: A Triple-A (Agent-Action-Aim) framework of self-efficacy beliefs in the context of collective social and ecological aims. *Personal. Soc. Psychol. Rev.* 28, 11–53 (2024). https://doi.org/10.1177/10888683231178056

8. Thunberg, G. *No One Is Too Small to Make a Difference.* (Penguin Books, 2019).

9. Besta, T., Jaśkiewicz, M., Kosakowska-Berezecka, N., Lawendowski, R. & Zawadzka, A. M. What do I gain from joining crowds? Does self-expansion help to explain the relationship between identity fusion, group efficacy and collective action? *Eur. J. Soc. Psychol.* 48, O152–O167 (2018). https://doi.org/10.1002/ejsp.2332

10. Brügger, A., Gubler, M., Steentjes, K. & Capstick, S. B. Social identity and risk perception explain participation in the Swiss youth climate strikes. *Sustainability* 12, 10605 (2020). https://doi.org/10.3390/su122410605

11. Rees, J. H. & Bamberg, S. Climate protection needs societal change: Determinants of intention to participate in collective climate action. *Eur. J. Soc. Psychol.* 44, 466–473 (2014). https://doi.org/10.1002/ejsp.2032

12. Landmann, H. & Rohmann, A. Being moved by protest: Collective efficacy beliefs and injustice appraisals enhance collective action intentions for forest protection via positive and negative emotions. *J. Environ. Psychol.* 71, 101491 (2020). https://doi.org/10.1016/j.jenvp.2020.101491

13. van Zomeren, M., Postmes, T. & Spears, R. On conviction's collective consequences: Integrating moral conviction with the social identity model of collective action: Conviction's collective consequences. *Br. J. Soc. Psychol.* 51, 52–71 (2012). https://doi.org/10.1111/j.2044-8309.2010.02000.x

14. Gulliver, R., Chapman, C. M., Solly, K. N. & Schultz, T. Testing the impact of images in environmental campaigns. *J. Environ. Psychol.* 71, 101468 (2020). https://doi.org/10.1016/j.jenvp.2020.101468

15. Doherty, K. L. & Webler, T. N. Social norms and efficacy beliefs drive the alarmed segment's public-sphere climate actions. *Nat. Clim. Change* 6, 879–884 (2016). https://doi.org/10.1038/nclimate3025

16. van Zomeren, M., Spears, R. & Leach, C. W. Experimental evidence for a dual pathway model analysis of coping with the climate crisis. *J. Environ. Psychol.* 30, 339–346 (2010). https://doi.org/10.1016/j.jenvp.2010.02.006

17. Sabherwal, A. *et al.* The Greta Thunberg Effect: Familiarity with Greta Thunberg predicts intentions to engage in climate activism in the United States. *J. Appl. Soc. Psychol.* 51, 321–333 (2021). https://doi.org/10.1111/jasp.12737

18. Thomas, E. F. & Louis, W. R. When will collective action be effective? Violent and non-violent protests differentially influence perceptions of legitimacy and efficacy among sympathizers. *Pers. Soc. Psychol. Bull.* 40, 263–276 (2014). https://doi.org/10.1177/0146167213510525

19. Carmona-Moya, B., Calvo-Salguero, A. & Aguilar-Luzón, M.-C. EIMECA: A proposal for a model of environmental collective action. *Sustainability* 13, 5935 (2021). https://doi.org/10.3390/su13115935

20. Wallis, H. & Loy, L. S. What drives pro-environmental activism of young people? A survey study on the Fridays For Future movement. *J. Environ. Psychol.* 74, 101581 (2021). https://doi.org/10.1016/j.jenvp.2021.101581

21. Cologna, V., Hoogendoorn, G. & Brick, C. To strike or not to strike? An investigation of the determinants of strike participation at the Fridays for Future climate strikes in Switzerland. *PLOS ONE* 16, e0257296 (2021). https://doi.org/10.1371/journal.pone.0257296

22. Thomas, E. F. *et al.* Whatever happened to Kony2012? Understanding a global internet phenomenon as an emergent social identity. *Eur. J. Soc. Psychol.* 45, 356–367 (2015). https://doi.org/10.1002/ejsp.2094

23. van Zomeren, M., Pauls, I. L. & Cohen-Chen, S. Is hope good for motivating collective action in the context of climate change? Differentiating hope's emotion- and problem-focused coping functions. *Glob. Environ. Change* 58, 101915 (2019). https://doi.org/10.1016/j.gloenvcha.2019.04.003

24. Hamann, K. R. S. & Reese, G. My influence on the world (of others): Goal efficacy beliefs and efficacy affect predict private, public, and activist pro-environmental behavior. *J. Soc. Issues* 76, 35–53 (2020). https://doi.org/10.1111/josi.12369

25. Bamberg, S., Rees, J. & Seebauer, S. Collective climate action: Determinants of participation intention in community-based pro-environmental initiatives. *J. Environ. Psychol.* 43, 155–165 (2015). https://doi.org/10.1016/j.jenvp.2015.06.006

26. Mazzoni, D., van Zomeren, M. & Cicognani, E. The motivating role of perceived right violation and efficacy beliefs in identification with the Italian water movement. *Polit. Psychol.* 36, 315–330 (2015). https://doi.org/10.1111/pops.12101

27. Kauhausen, K. *Creating a Climate for Change.* (Otto-von-Guericke-Universität, 2015).

28. Hamann, K. R. S., Holz, J. R. & Reese, G. Coaching for a sustainability transition: Empowering student-led sustainability initiatives by developing skills, group identification, and efficacy beliefs. *Front. Psychol.* 12, 623972 (2021). https://doi.org/10.3389/fpsyg.2021.623972

29. Hamann, K. R. S., von Agris, A.-S. & Markus, L. Investigating the predictors of collective action intensity and health. https://osf.io/c7vsn/ (2023).

30. Einwohner, R. L. Motivational framing and efficacy maintenance: Animal rights activists' use of four fortifying strategies. *Sociol. Q.* 43, 509–526 (2002). https://doi.org/10.1111/j.1533-8525.2002.tb00064.x

31. Drury, J., Cocking, C., Beale, J., Hanson, C. & Rapley, F. The phenomenology of empowerment in collective action. *Br. J. Soc. Psychol.* 44, 309–328 (2005). https://doi.org/10.1348/014466604X18523

32. Fernandez-Ballesteros, R., Diez-Nicolas, J., Caprara, G. V., Barbaranelli, C. & Bandura, A. Determinants and structural relation of personal efficacy to collective efficacy. *Appl. Psychol.* 51, 107–125 (2002). https://doi.org/10.1111/1464-0597.00081

33. Corcoran, K. E., Pettinicchio, D. & Young, J. T. N. The context of control: A cross-national investigation of the link between political institutions, efficacy, and collective action. *Br. J. Soc. Psychol.* 50, 575–605 (2011). https://doi.org/10.1111/j.2044-8309.2011.02076.x

34. Drury, J. & Reicher, S. Explaining enduring empowerment: A comparative study of collective action and psychological outcomes. *Eur. J. Soc. Psychol.* 35, 35–58 (2005). https://doi.org/10.1002/ejsp.231

35. Hornsey, M. J., Chapman, C. M. & Oelrichs, D. M. Ripple effects: Can information about the collective impact of individual actions boost perceived efficacy about

climate change? *J. Exp. Soc. Psychol.* 97, 104217 (2021). https://doi.org/10.1016/j.jesp.2021.104217

36. Xue, W. *et al.* Combining threat and efficacy messaging to increase public engagement with climate change in Beijing, China. *Clim. Change* 137, 43–55 (2016). https://doi.org/10.1007/s10584-016-1678-1

37. Jugert, P. *et al.* Collective efficacy increases pro-environmental intentions through increasing self-efficacy. *J. Environ. Psychol.* 48, 12–23 (2016). https://doi.org/10.1016/j.jenvp.2016.08.003

38. Ende Gelände. Wir sind das Investitionsrisiko! Ende Gelände! 13.–16. Mai 2016 in der Lausitz. https://www.youtube.com/watch?v=jMKCVPONww0 (2016).

39. Morton, T. A., Rabinovich, A., Marshall, D. & Bretschneider, P. The future that may (or may not) come: How framing changes responses to uncertainty in climate change communications. *Glob. Environ. Change* 21, 103–109 (2011). https://doi.org/10.1016/j.gloenvcha.2010.09.013

40. Perspective Daily – dein Online-Magazin für mehr Überblick. *Perspective Daily.* https://perspective-daily.de

41. Müller, T. Kämpfen und zweifeln. *nd-aktuell.de.* https://www.nd-aktuell.de/artikel/1147613.klimagerechtigkeit-kaempfen-und-zweifeln.html (2021).

42. Tausch, N. & Becker, J. C. Emotional reactions to success and failure of collective action as predictors of future action intentions: A longitudinal investigation in the context of student protests in Germany. *Br. J. Soc. Psychol.* 52, 525–542 (2013). https://doi.org/10.1111/j.2044-8309.2012.02109.x

43. Spiegel Wissenschaft. So tickt Greta Thunbergs Klimabewegung. *Der Spiegel.* https://www.spiegel.de/wissenschaft/mensch/greta-thunberg-wissenschaftler-untersuchen-fridays-for-future-bewegung-a-1282826.html (2019).

44. Graham-Harrison, E. Greta Thunberg takes climate fight to Germany's threatened Hambach Forest. *The Guardian.* https://www.theguardian.com/environment/2019/aug/10/greta-thunberg-climate-change-fight-germany-hambach-forest (2019).

45. Reznickova, A. & Zepeda, L. Can self-determination theory explain the self-perpetuation of social innovations? A case study of slow food at the University of Wisconsin – Madison. *J. Community Appl. Soc. Psychol.* 26, 3–17 (2016). https://doi.org/10.1002/casp.2229

46. Almers, E. Pathways to action competence for sustainability—Six themes. *J. Environ. Educ.* 44, 116–127 (2013). https://doi.org/10.1080/00958964.2012.719939

47. Howell, A. J. Self-affirmation theory and the science of well-being. *J. Happiness Stud.* 18, 293–311 (2017). https://doi.org/10.1007/s10902-016-9713-5

48. Drury, J. & Reicher, S. The intergroup dynamics of collective empowerment: Substantiating the social identity model of crowd behavior. *Group Process. Intergroup Relat.* 2, 381–402 (1999). https://doi.org/10.1177/1368430299024005

49. Wang, E. S.-T. & Lin, H.-C. Sustainable development: The effects of social normative beliefs on environmental behaviour. *Sustain. Dev.* 25, 595–609 (2017). https://doi.org/10.1002/sd.1680

50. Wang, X. The role of attitudinal motivations and collective efficacy on Chinese consumers' intentions to engage in personal behaviors to mitigate climate change. *J. Soc. Psychol.* 158, 51–63 (2018). https://doi.org/10.1080/00224545.2017.1302401

51. Madman Films. Tomorrow – Official Trailer. https://www.youtube.com/watch?v=0SI-Kyam_Jk (2016).

52. Swim, J. K., Geiger, N. & Lengieza, M. L. Climate change marches as motivators for bystander collective action. *Front. Commun.* 4, 4 (2019). https://doi.org/10.3389/fcomm.2019.00004

53. van Zomeren, M., Postmes, T. & Spears, R. Toward an integrative social identity model of collective action: A quantitative research synthesis of three socio-psychological perspectives. *Psychol. Bull.* 134, 504–535 (2008). https://doi.org/10.1037/0033-2909.134.4.504

54. Bieniek-Tobasco, A. *et al.* Communicating climate change through documentary film: Imagery, emotion, and efficacy. *Clim. Change* 154, 1–18 (2019). https://doi.org/10.1007/s10584-019-02408-7

55. Olson, M. *The Logic of Collective Action: Public Goods and the Theory of Groups.* (Harvard University Press, 1971). https://doi.org/10.2307/j.ctvjsf3ts

56. Hornsey, M. J., Chapman, C. M. & Oelrichs, D. M. Why it is so hard to teach people they can make a difference: Climate change efficacy as a non-analytic form of reasoning. *Think. Reason.* 1–19 (2022). https://doi.org/10.1080/13546783.2021.1893222

57. Antonetti, P. & Maklan, S. Feelings that make a difference: How guilt and pride convince consumers of the effectiveness of sustainable consumption choices. *J. Bus. Ethics* 124, 117–134 (2014). https://doi.org/10.1007/s10551-013-1841-9

58. Landmann, H. & Naumann, J. Being positively moved by climate protest predicts peaceful collective action. *Glob. Environ. Psychol.* https://www.psycharchives.org/en/item/72fb35c7-7174-47b6-985c-7d1a477965db (2023).

59. Hornsey, M. J. *et al.* Evidence for motivated control: Understanding the paradoxical link between threat and efficacy beliefs about climate change. *J. Environ. Psychol.* 42, 57–65 (2015). https://doi.org/10.1016/j.jenvp.2015.02.003

60. Valentino, N. A., Gregorowicz, K. & Groenendyk, E. W. Efficacy, emotions and the habit of participation. *Polit. Behav.* 31, 307–330 (2009). https://doi.org/10.1007/s11109-008-9076-7

61. Cohen-Chen, S. & Van Zomeren, M. Yes we can? Group efficacy beliefs predict collective action, but only when hope is high. *J. Exp. Soc. Psychol.* 77, 50–59 (2018). https://doi.org/10.1016/j.jesp.2018.03.016

62. Ojala, M. Hope in the face of climate change: Associations with environmental engagement and student perceptions of teachers' emotion communication style and future orientation. *J. Environ. Educ.* 46, 133–148 (2015). https://doi.org/10.1080/00958964.2015.1021662

63. Marlon, J. R. *et al.* How hope and doubt affect climate change mobilization. *Front. Commun.* 4, 20 (2019). https://doi.org/10.3389/fcomm.2019.00020

64. Brosch, T. Affect and emotions as drivers of climate change perception and action: A review. *Curr. Opin. Behav. Sci.* 42, 15–21 (2021). https://doi.org/10.1016/j.cobeha.2021.02.001

65. Hornsey, M. J. & Fielding, K. S. A cautionary note about messages of hope: Focusing on progress in reducing carbon emissions weakens mitigation motivation. *Glob. Environ. Change* 39, 26–34 (2016). https://doi.org/10.1016/j.gloenvcha.2016.04.003

66. Greenaway, K. H., Cichocka, A., van Veelen, R., Likki, T. & Branscombe, N. R. Feeling hopeful inspires support for social change. *Polit. Psychol.* 37, 89–107 (2016). https://doi.org/10.1111/pops.12225

67. Shuman, E., Cohen-Chen, S., Hirsch-Hoefler, S. & Halperin, E. Explaining normative versus nonnormative action: The role of implicit theories. *Polit. Psychol.* 37, 835–852 (2016). https://doi.org/10.1111/pops.12325

68. Jiménez-Moya, G., Rodríguez-Bailón, R., Spears, R. & de Lemus, S. Collective resistance despite complicity: High identifiers rise above the legitimization of disadvantage by the in-group. *Br. J. Soc. Psychol.* 56, 103–124 (2017). https://doi.org/10.1111/bjso.12182

69. Moser, S. & Seebauer, S. Has the COVID-19 pandemic strengthened confidence in managing the climate crisis? Transfer of efficacy beliefs after experiencing lockdowns

in Switzerland and Austria. *Front. Psychol.* 13, 892735 (2022). https://doi.org/10.3389/fpsyg.2022.892735

70. Obermiller, C. The baby is sick/The baby is well: A test of environmental communication appeals. *J. Advert.* 24, 55–70 (1995). https://doi.org/10.1080/00913367.1995.10673476

71. Van de Velde, L., Verbeke, W., Popp, M. & Van Huylenbroeck, G. The importance of message framing for providing information about sustainability and environmental aspects of energy. *Energy Policy* 38, 5541–5549 (2010). https://doi.org/10.1016/j.enpol.2010.04.053

72. Grant, A. M. Making positive change: A randomized study comparing solution-focused vs. problem-focused coaching questions. *J. Syst. Ther.* 31, 21–35 (2012). https://doi.org/10.1521/jsyt.2012.31.2.21

73. Bosone, L., Thiriot, S., Chevrier, M., Rocci, A. & Zenasni, F. Visioning sustainable futures: Exposure to positive visions increases individual and collective intention to act for a decarbonated world. *Global Environmental Psychology* 2 (2024). https://doi.org/10.5964/gep.11105

74. Fernando, J. W., O'Brien, L. V., Burden, N. J., Judge, M. & Kashima, Y. Greens or space invaders: Prominent utopian themes and effects on social change motivation. *Eur. J. Soc. Psychol.* 50, 278–291 (2020). https://doi.org/10.1002/ejsp.2607

75. Wright, J. D., Schmitt, M. T., Mackay, C. M. L. & Neufeld, S. D. Imagining a sustainable world: Measuring cognitive alternatives to the environmental status quo. *J. Environ. Psychol.* 72, 101523 (2020). https://doi.org/10.1016/j.jenvp.2020.101523

76. Wright, J. D., Schmitt, M. T. & Mackay, C. M. L. Access to environmental cognitive alternatives predicts pro-environmental activist behavior. *Environ. Behav.* 31 (2022). https://doi.org/10.1177/00139165211065008

77. Fernando, J. W. *et al.* Functions of utopia: How utopian thinking motivates societal engagement. *Pers. Soc. Psychol. Bull.* 44, 779–792 (2018). https://doi.org/10.1177/0146167217748604

78. Barr, D. & Drury, J. Activist identity as a motivational resource: Dynamics of (Dis)empowerment at the G8 Direct Actions, Gleneagles, 2005. *Soc. Mov. Stud.* 8, 243–260 (2009). https://doi.org/10.1080/14742830903024333

79. Bleh, J. Social experiences as inspiration for societal imagination: First insights from three field studies (presented at ICEP, 2023).

80. Bain, P. G., Hornsey, M. J., Bongiorno, R., Kashima, Y. & Crimston, C. R. Collective futures: How projections about the future of society are related to actions and attitudes supporting social change. *Pers. Soc. Psychol. Bull.* 39, 523–539 (2013). https://doi.org/10.1177/0146167213478200

81. Hamann, K. How to envision an ecological future? An experimental study on the effectiveness of presented vs. self-generated visions (presented at DGPs-Kongress, 2022).

82. Fritsche, I., Barth, M., Jugert, P., Masson, T. & Reese, G. A Social Identity Model of Pro-Environmental Action (SIMPEA). *Psychol. Rev.* 125, 245–269 (2018). https://doi.org/10.1037/rev0000090

83. Thomas, E. F. & McGarty, C. A. The role of efficacy and moral outrage norms in creating the potential for international development activism through group-based interaction. *Br. J. Soc. Psychol.* 48, 115–134 (2009). https://doi.org/10.1348/014466608X313774

84. Scafuto, F. & La Barbera, F. Protest against waste contamination in the 'Land of Fires': Psychological antecedents for activists and non-activists. *J. Community Appl. Soc. Psychol.* 26, 481–495 (2016). https://doi.org/10.1002/casp.2275

85. Collado, S. & Evans, G. W. Outcome expectancy: A key factor to understanding childhood exposure to nature and children's pro-environmental behavior. *J. Environ. Psychol.* 61, 30–36 (2019). https://doi.org/10.1016/j.jenvp.2018.12.001

86. Kerr, N. L. Illusions of efficacy: The effects of group size on perceived efficacy in social dilemmas. *J. Exp. Soc. Psychol.* 25, 287–313 (1989). https://doi.org/10.1016/0022-1031(89)90024-3

87. Hamann, K., Masson, T., Fritsche, I., Dasch, S., von der Kaus, K. *Report on experimental lab studies on energy citizenship – energy community set-ups, energy visions and collective agency as predictors of energy citizenship and pro-environmental spillover.* https://ec2proj ect.eu/resources/downloads (2023).

88. Lickel, B. *et al.* Varieties of groups and the perception of group entitativity. *J. Pers. Soc. Psychol.* 78, 223–246 (2000). https://doi.org/10.1037/0022-3514.78.2.223

89. IPU Initiative Psychologie im Umweltschutz. https://ipu-ev.de/

90. Climate Outreach. https://climateoutreach.org/

91. Fritsche, I. Agency through the we: Group-based control theory. *Curr. Dir. Psychol. Sci.* 31, 194–201 (2022). https://doi.org/10.1177/09637214211068838

92. Fritsche, I. & Masson, T. Collective climate action: When do people turn into collective environmental agents? *Curr. Opin. Psychol.* 42, 114–119 (2021). https://doi.org/10.1016/j.copsyc.2021.05.001

93. Cocking, C. & Drury, J. Generalization of efficacy as a function of collective action and intergroup relations: Involvement in an anti-roads struggle. *J. Appl. Soc. Psychol.* 34, 417–444 (2004). https://doi.org/10.1111/j.1559-1816.2004.tb02555.x

94. Drury, J., Evripidou, A. & van Zomeren, M. Empowerment: The Intersection of Identity and Power in Collective Action. In *Power and Identity* (eds. Sindic, D., Barreto, M. & Costa-Lopes, R.) 94–116 (Psychology Press, 2015).

95. Hornsey, M. J. *et al.* Why do people engage in collective action? Revisiting the role of perceived effectiveness. *J. Appl. Soc. Psychol.* 36, 1701–1722 (2006). https://doi.org/10.1111/j.0021-9029.2006.00077.x

96. Harré, N. Community service or activism as an identity project for youth. *J. Community Psychol.* 35, 711–724 (2007). https://doi.org/10.1002/jcop.20174

97. Truelove, H. B., Carrico, A. R., Weber, E. U., Raimi, K. T. & Vandenbergh, M. P. Positive and negative spillover of pro-environmental behavior: An integrative review and theoretical framework. *Glob. Environ. Change* 29, 127–138 (2014). https://doi.org/10.1016/j.gloenvcha.2014.09.004

98. Lee, D. S. & Ybarra, O. Cultivating effective social support through abstraction: Reframing social support promotes goal-pursuit. *Pers. Soc. Psychol. Bull.* 43, 453–464 (2017). https://doi.org/10.1177/0146167216688205

99. Kruglanski, A. W. *et al.* A Theory of Goal Systems. In *Advances in Experimental Social Psychology* (ed. Zanna, M. P.) Vol. 34, 331–378 (Academic Press, 2002). https://doi.org/10.1016/S0065-2601(02)80008-9

100. Lee, F. L. F. The perceptual bases of collective efficacy and protest participation: The case of pro-democracy protests in Hong Kong. *Int. J. Public Opin. Res.* 22, 392–411 (2010). https://doi.org/10.1093/ijpor/edq023

101. Tausch, N. *et al.* Explaining radical group behavior: Developing emotion and efficacy routes to normative and nonnormative collective action. *J. Pers. Soc. Psychol.* 101, 129–148 (2011). https://doi.org/10.1037/a0022728

102. Osborne, D., Yogeeswaran, K. & Sibley, C. G. Hidden consequences of political efficacy: Testing an efficacy–apathy model of political mobilization. *Cultur. Divers. Ethnic Minor. Psychol.* 21, 533–540 (2015). https://doi.org/10.1037/cdp0000029

103. Gulevich, O., Sarieva, I., Nevruev, A. & Yagiyayev, I. How do social beliefs affect political action motivation? The cases of Russia and Ukraine. *Group Process. Intergroup Relat.* 20, 382–395 (2017). https://doi.org/10.1177/1368430216683531

104. netzwerk n e.V. Bildungsmaterialien / netzwerk n. https://netzwerk-n.org

105. Jung, D. I. & Sosik, J. J. Transformational leadership in work groups: The role of empowerment, cohesiveness, and collective-efficacy on perceived group performance. *Small Group Res.* 33, 313–336 (2002). https://doi.org/10.1177/10496402033003002

106. Reese, G. & Junge, E. Keep on rockin' in a (plastic-)free world: Collective efficacy and pro-environmental intentions as a function of task difficulty. *Sustainability* 9, 200 (2017). https://doi.org/10.3390/su9020200

107. Kantor, D. & Lehr, W. *Inside the Family*. (San Francisco: Jossey-Bass, 1975).

108. Cleff, A. Helping team members stretch their communication muscles: Kantor four player model. *Andy Cleff.* https://www.andycleff.com/2021/04/stretch-team-communication-muscles-kantor-four-player-model/

109. Commons Librarian. Bill Moyer's movement action plan and four roles of activism. *The Commons.* https://commonslibrary.org/bill-moyers-movement-action-plan-and-four-roles-of-activism/ (2022).

110. Adams, N. *Ecologies of UK Social Movements*. https://thinkingdoingchanging.files.wordpress.com/2019/05/the-ecologies-of-uk-social-movements-.pdf (2019).

111. Hoffman, A. J. Shades of green. *Stanf. Soc. Innov. Rev.* 7, 4049 (2009).

112. Heitfeld, M. & Reif, A. Transformation gestalten lernen.https://www.germanwatch.org/de/19607 (2021).

113. Scientists for Future. About. *Scientists 4 Future* https://scientists4future.org/

114. Tassone, V. C., Dik, G. & van Lingen, T. A. Empowerment for sustainability in higher education through the EYE learning tool. *Int. J. Sustain. High. Educ.* 18, 341–358 (2017). https://doi.org/10.1108/IJSHE-12-2015-0209

115. Vestergren, S., Drury, J. & Chiriac, E. H. The biographical consequences of protest and activism: A systematic review and a new typology. *Soc. Mov. Stud.* 16, 203–221 (2017). https://doi.org/10.1080/14742837.2016.1252665

116. Cable, S. Women's social movement involvement: The role of structural availability in recruitment and participation processes. *Sociol. Q.* 33, 35–50 (1992). https://doi.org/10.1111/j.1533-8525.1992.tb00362.x

117. Siy, J. O. *et al.* Does the follow-your-passions ideology cause greater academic and occupational gender disparities than other cultural ideologies? *J. Pers. Soc. Psychol.* 125, 548–570 (2023). https://doi.org/10.1037/pspi0000421

118. Peterson, N. A. & Zimmerman, M. A. Beyond the individual: Toward a nomological network of organizational empowerment. *Am. J. Community Psychol.* 34, 129–145 (2004). https://doi.org/10.1023/B:AJCP.0000040151.77047.58

6

PSYCHOLOGICAL EFFECTS OF COLLECTIVE CLIMATE ACTION

ACTION FIRST, MOTIVATION SECOND

After reading these chapters, you may be thinking, *okay, so a group first needs to motivate people and create and strengthen feelings of belonging, injustice, and efficacy, and then they will join collective climate actions*. This is somewhat true, but it would be a false conclusion to think that climate action groups always have to start with these motivations.

DOI: 10.4324/9781003558439-7

It's important to keep in mind that the relationships between concepts used within the field of psychology are often tested in cross-sectional settings, which means several concepts can wind up being analyzed at the same time. This muddies the water when it comes to what psychology researchers can predict or assert. For example, it is difficult to determine whether people feel angry and then decide to join a collective action, or whether they join a collective action, are confronted with unjust treatment, and begin to feel angry. The causality remains unclear.

Building climate courage could also work this way: participating in climate action can breed motivation[1]. Phrased as an active strategy, this means "Action first, motivation second". For example, an individual may bring their roommate to an anti-nuclear-power protest. An experience like this could evoke feelings of efficacy in the roommate, which could, alongside additional experiences, motivate them to pursue a political career. Indeed, research shows that being asked to join a protest is a primary motivational source.[2] As well, our first protest experience and how we perceive it can be a defining moment.[3]

The previously described study on the *Water for Life* movement also shows that this causal direction is possible. This study involved participants in a planning session on how to support the movement, which in itself represents collective climate action. Participating in this session was found to increase social identification and efficacy beliefs.[4] Studies have also found that political participation (action) can promote political efficacy beliefs (motivation) in this area even after political defeat.[5]

The "action first, motivation second" idea highlights that an action such as bringing friends and family to a collective climate action might be a powerful starting point for motivation. One of our authors recalls a time her friend treated her to brunch at a vegan café. Alongside the delicious food, people were passing around a petition on animal rights, something that really surprised her. This took the brunch from being a vegan lifestyle event to being an enjoyable political networking opportunity.

"Action first, motivation second" might be an especially valuable idea for climate educators. Instead of convincing students of injustices and that their actions can make a difference, tasks can be designed such that they already amount to collective climate action. For example, instead of having students read a text about climate injustice, have them draw up a petition together on a climate-related topic of their choice. This way, they might begin to feel like a climate action group addressing a morally relevant issue and potentially receive positive feedback that builds efficacy.

The effect that collective climate action can have on motivation is likely to depend on whether we have a good or bad experience with it; for example, in terms of whether our needs for belonging and meaning are satisfied or intensified and whether we perceive the action as a success or as a failure.

Box 6.1: The bottom line

Collective climate action can influence an individual's social identification, moral beliefs, and efficacy beliefs. It is therefore important that we do not underestimate the effect of bringing newcomers directly into politicized contexts where they are likely to have transformational experiences.

Box 6.2: Food for thought – The intention-behavior gap of activism

In environmental psychology, one of the big questions is how to bridge the intention-behavior gap.[6] An example of this gap would be wanting to start living more sustainably in the private sphere but it not transferring into action. Recent unpublished research on the climate action group *Extinction Rebellion* suggests that the same seems to be true for collective climate action – many people who report intentions to join actions never actually do so.[7]

While the Author Team knows of no research on bridging the intention-behavior gap for collective climate action, it is likely that existing behavior change strategies from the private sphere can be applicable. One such strategy could involve the use of visual reminders to keep our goals always in our sight. For example, an individual wanting to attend the local protest for a bicycle-friendly neighborhood might benefit from putting a postcard with a bicycle on it somewhere they're sure to see it every time they leave the house.

Another, more elaborate strategy would be to form implementation intentions by asking ourselves when, where, and with whom we want to perform a new behavior (for example, attending the environmental student group meetings) and by anticipating potential barriers that might get in our way (for example, exam week).

A detailed description of these strategies can be found in the book *Psychology of Environmental Protection*.[6]

EFFECTS OF AN ACTION'S SUCCESS OR FAILURE

The previous chapter on efficacy beliefs already tapped into the relevance of success and failure. Let's now explore the options people have when it comes to reacting to a successful or unsuccessful collective climate action.

Winnifred Louis, professor at the University of Queensland, and colleagues propose that a successful action would lead people to choose similar strategies the next time.[8] If a climate action group blocks an intersection in order to highlight the need for sustainable mobility infrastructure and they receive positive feed-back (e.g., favorable media coverage), they may want to try a similar type of action next time. From our experience within the Author Team, it could also be possible that this successful action serves as the stable foundation on which other, slightly differing forms of action can be built – through a strengthened sense of identity, moral beliefs, and efficacy.

If a collective climate action was not successful, measured perhaps by a lack of media coverage, public debate highlighting the action's flaws, or violent police interaction, group members may have conflicting opinions as to what they believe the group should do next. As described, this reaction may depend on whether a sense of efficacy in the group, the action or the government has been lost.[9,10] In response to an unsuccessful collective action, some group members may want to choose a more radical form of action next time[8] – one that they believe will be more effective in raising attention. If the group as a whole is unwilling to change in this direction, a radical flank may emerge. Other group members might think that they should indeed change their strategy but use a less radical form in order to garner more public support. Others still might want to stick to the same strategy but double their efforts and plan it in a way that makes the moral urgency even more prominent,[8] thereby changing its framing and features. And then there are those who, in the wake of failure, lose their connection to and identification with the group entirely.[8]

While the ideas by Louis and colleagues are still being tested,[8,11] they already point to two interesting implications. On the one hand, failure has the potential to divide groups, not only because people leave but also because people draw differing conclusions from this failure. In other words: people choose different strategies to regain their sense of efficacy, some of which may be incompatible with those chosen by other group members. On the other hand, if a group experiences success and feels effective "enough", that success may also lead to them staying in their comfort zone.

 Box 6.3: The bottom line

After a failure, some members may need to regain a sense of their social identification to feel comfortable with the group (again), such that they can then discuss future, potentially more effective, collective climate actions with a clear head and compassionate mindset.

STRATEGIES FOR COPING WITH FAILURE

There are certain strategies that might help us cope with failure and disap-pointment. This section delves into a few strategies taken from the real-world experiences of the Author Team, interviews with environmentally engaged people about their efficacy beliefs, and further ideas in this book.

Social strategies (a social and shared mindset)

As the climate crisis is a collective problem, it's best not to try to deal with failure and any accompanying feelings alone. Often times when we try to deal with failure on our own we end up losing some or all of our motivation. This is why coping with failure should be a collective challenge. So, let's look at some strategies we can employ with others:

- Look for people who cheer you up with their positivity and support
- Create opportunities to process the failure together – grieving together might provide some relief
- Seek professional help (counseling, therapy)
- Try not to individualize a problem too much when it is actually a socio-structural problem – you are not suffering alone
- If you are part of a climate action group, interact with other members and engage with the coping strategies described in the following

Reframing strategies (another perspective)

There are some mental strategies that can be used to develop different perspectives on a situation and help us see positive aspects where we previously saw only failure.

- Reframe the goals you wanted to achieve: the campaign did not receive a lot of attention, but it did foster group cohesion.
- Highlight small successes: the campaign did not receive a lot of attention, but it was still a first step in the right direction – people now know we exist.
- Emphasize the bigger picture: the campaign did not receive a lot of attention, but we know activism is a marathon (or even hike), not a sprint.
- Devalue the importance of a winning streak: the campaign did not receive a lot of attention, but let's remember that it is always a trial-and-error process. Sometimes you win, sometimes you lose.

Distancing strategies (another focus)

The positive effects of distraction are often neglected by people engaged in collective climate action, but doing things that bring us joy is critical in helping us recover from moments of failure. Though joy is highly individual, some of the things that might cheer us up are going to a festival, seeing a play, spending time with our family or friends, going for a swim, making art, or simply spending a relaxing day at the park.

Changing strategies (another action)

There is always the possibility that a failure will make us feel like we should just stop doing what we are doing and switch to another group or a different kind of climate action. Even when we feel strongly obligated to a group, this feeling is nothing to be ashamed of. Indeed, this may be the right course of action in a given situation. It is true that if a lot of people move quickly from one action or group to another, the overall resilience and continuity of the climate movement might suffer. However, it is also true that if, over time, we notice that none of the mentioned strategies help us cope, and our needs for belonging, meaning, and efficacy are not being met, it may be exactly the right thing to do to change to another kind of collective climate action. For example, this might be moving from large-scale climate protests to a community-supported agriculture project.

Another (possibly even more resilient) option is to change your group's strategy collectively, maybe in a joint strategy meeting with other members or with other climate action groups. See Chapter 10 for guidance on how to (re)set goals collectively.

THE SPILLOVER EFFECT BETWEEN PRIVATE BEHAVIORS AND COLLECTIVE ACTION

In the field of psychology, there has been an ongoing debate over how different types of climate action influence each other – how they spill over into one another (spillover effect). For example, if a person is starting to commute to work by bike instead of by car, how likely is it that behavior will spill over into motivating them to attend the next demonstration against the car industry?

There are two competing hypotheses for this phenomenon: the gateway hypothesis and the getaway hypothesis.[12] The gateway hypothesis suggests that private actions such as adopting a vegan diet might be a gateway to collective action, such as joining an animal rights group. This would be a sign of positive spillover. The getaway hypothesis states that people might feel they have done their duty in adopting a vegan diet and thus don't need to join an animal rights group as they are already active in their private life. This perception of a moral license leads to negative spillover.

One study from Stockholm University and the Institute of Political Science Louvain-Europe provided empirical support for the gateway hypothesis, showing a small positive spillover effect from private behavior to collective action.[12] However, an experimental study and our own unpublished longitudinal work were unable to replicate this finding with other designs.[13,14] What our unpublished work did find is that if there is positive spillover at all, it is more often that collective action spills over into private actions. This spillover from collective to private action is also what a lot of qualitative interviews demonstrate.[15-18] For example, in interviews with members of *Sea Shepherd*, an environmental action group that uses diverse tactics (also civil disobedience) to protect marine ecosystems, one interviewee reported a positive spillover effect.[17] After joining the organization, they spotted the documentary *Earthlings* lying on a table and watched it. This documentary gave them the urge to refrain from eating animal products and have a healthier lifestyle. A spillover may be especially likely when these private actions are, in some notable way, similar to and exist within the same domain as the collective actions an individual engages in.[19]

Overall, these findings show that climate action groups should not rely on spillover to just happen. Changes in private and collective action can be built on slightly different motivational grounds, such that changing one does not necessarily change the other.

> If groups want to change socio-political structures and want to encourage people to engage in collective climate action, they are advised not to concentrate on changing private actions first.

Let's take this example: a transition town group wants to build an alternative economy. As this is a big task, they start searching for ways to begin on a small scale. One group member proposes the idea of running a sustainable city tour so the public can learn where they can buy fairtrade products, as they believe this might result in people eventually becoming motivated to really change economic structures. Yet, this type of action will not necessarily cause the desired spillover effect and might just take up valuable time and resources. If this transition town group wants to promote collective climate action, they would be better off talking about how to collectively build an alternative economy. At best, such a tour might end with a collective climate action by the participants, such as signing a fairtrade petition, sending a letter to a relevant public official, or creating a joint painting on the topic that will later be displayed in public space.

 Box 6.4: The bottom line

Research reveals that private and collective climate actions do not necessarily spill over into one another – there is no guarantee of a spillover effect occurring. If a group wants to promote collective climate action, it should aim for exactly this action and not get mired in changing private behaviors.

References

1. Vestergren, S., Drury, J. & Chiriac, E. H. The biographical consequences of protest and activism: A systematic review and a new typology. *Soc. Mov. Stud.* 16, 203–221 (2017). https://doi.org/10.1080/14742837.2016.1252665
2. Schussman, A. & Soule, S. A. Process and protest: Accounting for individual protest participation. *Soc. Forces* 84, 1083–1108 (2005). https://doi.org/10.1353/sof.2006.0034
3. Verhulst, J. & Walgrave, S. The first time is the hardest? A cross-national and cross-issue comparison of first-time protest participants. *Polit. Behav.* 31, 455–484 (2009). https://doi.org/10.1007/s11109-008-9083-8
4. Thomas, E. F. & McGarty, C. A. The role of efficacy and moral outrage norms in creating the potential for international development activism through group-based interaction. *Br. J. Soc. Psychol.* 48, 115–134 (2009). https://doi.org/10.1348/014466608 X313774

5. Valentino, N. A., Gregorowicz, K. & Groenendyk, E. W. Efficacy, emotions and the habit of participation. *Polit. Behav.* 31, 307–330 (2009). https://doi.org/10.1007/s11109-008-9076-7

6. Hamann, K., Löschinger, D. & Baumann, A. *Psychology of Environmental Protection – Handbook for Encouraging Sustainable Actions.* (2016) www.wandel-werk.org/en/materialien

7. Parkes, I., Thomas-Walters, L., Sabherwal, A., O'Dell, D. & Shreedhar, G. Social identification predicts behavioural engagement with environmental activist movements, but does not moderate the collective climate action intention-behaviour gap. SocArXiv. https://doi.org/10.31235/osf.io/4ksgw (2023).

8. Louis, W. *et al.* The volatility of collective action: Theoretical analysis and empirical data. *Polit. Psychol.* 41, 35–74 (2020). https://doi.org/10.1111/pops.12671

9. Gulevich, O., Sarieva, I., Nevruev, A. & Yagiyayev, I. How do social beliefs affect political action motivation? The cases of Russia and Ukraine. *Group Process. Intergroup Relat.* 20, 382–395 (2017). https://doi.org/10.1177/1368430216683531

10. Tausch, N. *et al.* Explaining radical group behavior: Developing emotion and efficacy routes to normative and nonnormative collective action. *J. Pers. Soc. Psychol.* 101, 129–148 (2011). https://doi.org/10.1037/a0022728

11. Lizzio-Wilson, M. *et al.* How collective-action failure shapes group heterogeneity and engagement in conventional and radical action over time. *Psychol. Sci.* 32, 519–535 (2021). https://doi.org/10.1177/0956797620970562

12. de Moor, J. & Verhaegen, S. Gateway or getaway? Testing the link between lifestyle politics and other modes of political participation. *Eur. Polit. Sci. Rev.* 12, 91–111 (2020). https://doi.org/10.1017/S1755773919000377

13. Lacroix, K. *et al.* Does personal climate change mitigation behavior influence collective behavior? Experimental evidence of no spillover in the United States. *Energy Res. Soc. Sci.* 94, 102875 (2022). https://doi.org/10.1016/j.erss.2022.102875

14. Hamann, K. R. S. *Psychological Empowerment in the Context of Environmental Protection: How Can Personal, Collective, and Participative Efficacy Beliefs Foster Pro-environmental Behavior and Activism?* (University Koblenz-Landau, 2022).

15. Cocking, C. & Drury, J. Generalization of efficacy as a function of collective action and intergroup relations: Involvement in an anti-roads struggle. *J. Appl. Soc. Psychol.* 34, 417–444 (2004). https://doi.org/10.1111/j.1559-1816.2004.tb02555.x

16. Drury, J. & Reicher, S. Explaining enduring empowerment: A comparative study of collective action and psychological outcomes. *Eur. J. Soc. Psychol.* 35, 35–58 (2005). https://doi.org/10.1002/ejsp.231

17. Stuart, A., Thomas, E. F., Donaghue, N. & Russell, A. "We may be pirates, but we are not protesters": Identity in the Sea Shepherd Conservation Society. *Polit. Psychol.* 34, 753–777 (2013). https://doi.org/10.1111/pops.12016

18. Vestergren, S., Drury, J. & Chiriac, E. H. How collective action produces psychological change and how that change endures over time: A case study of an environmental campaign. *Br. J. Soc. Psychol.* 57, 855–877 (2018). https://doi.org/10.1111/bjso.12270

19. Maki, A. *et al.* Meta-analysis of pro-environmental behaviour spillover. *Nat. Sustain.* 2, 307–315 (2019). https://doi.org/10.1038/s41893-019-0263-9

7 SUMMARY **OF THE** MODEL

DOI: 10.4324/9781003558439-8

APPLYING THE PILLARS OF MOTIVATION TO DRIVE COLLECTIVE CLIMATE ACTION

At this stage, you've gained a good overview of the psychology of how to motivate people to participate in collective climate action and how to build climate courage. Social identification, moral beliefs, and efficacy beliefs are the pillars of this courage. It is important to note that these three psychological concepts are strongly interrelated and can also influence each other.

The SIMCA model that formed the starting point for this book proposes that social identification influences perceptions of injustice and efficacy beliefs, which in turn leads to collective action.[1] It also assumes that social identification has a direct influence on collective action. Other research suggests that, especially in the formation of an action group, one's own and the group's efficacy beliefs and injustice perceptions are crucial to the development of a group identity, which in turn promotes collective action.[2–4] This means that, even if an action focuses on increasing one motivation, there is the chance that other motivations are also positively affected.

The Author Team has benefitted from planning and brainstorming using pre-structured questions, which are provided as a worksheet or canvas in Figure 7.1. This canvas will help you to apply the information presented in this book to a specific collective climate action. Before filling out the canvas, be sure to clearly define your target group (demographics, job, living environment, competences, resources), your main goal for the action, and the type of collective climate action you want to focus on. The task is simple: just write as many ideas as you want into the blank spaces, working individually or with others. In a second step, circle the more interesting ideas and discuss them. A printable version of this canvas can be downloaded from the *Wandelwerk*[5] website.

References

1. van Zomeren, M., Postmes, T. & Spears, R. Toward an integrative social identity model of collective action: A quantitative research synthesis of three socio-psychological perspectives. *Psychol. Bull.* 134, 504–535 (2008). https://doi.org/10.1037/0033-2909.134.4.504

2. Thomas, E. F., McGarty, C. & Mavor, K. I. Aligning identities, emotions, and beliefs to create commitment to sustainable social and political action. *Personal. Soc. Psychol. Rev.* 13, 194–218 (2009). https://doi.org/10.1177/1088868309341563

3. Thomas, E. F. & McGarty, C. A. The role of efficacy and moral outrage norms in creating the potential for international development activism through group-based interaction. *Br. J. Soc. Psychol.* 48, 115–134 (2009). https://doi.org/10.1348/014466608X313774

4. Thomas, E. F., Mavor, K. I. & McGarty, C. Social identities facilitate and encapsulate action-relevant constructs: A test of the social identity model of collective action. *Group Process. Intergroup Relat.* 15, 75–88 (2012). https://doi.org/10.1177/1368430211413619

5. Wandelwerk e.V. Wandelwerk Umweltpsychologie: Wir bringen Psychologie in den Umweltschutz. https://www.wandel-werk.org/.

Social identity	Moral beliefs	Efficacy beliefs
Which social categories and groups does the target group already identify with? How can we acknowledge these groups and their values in our communication?	Does our target group already perceive the injustices that we want to draw attention to? Are they angry enough?	Does the target group believe that the collective climate action can actually achieve something? Do they feel like they can contribute to the group?
Does the target group already identify with a climate action group? How can we increase identification with the help of social norms?	How do we want our collective climate action to be perceived in public? How can we strike a balance between public attention and support?	How can we strengthen a sense of efficacy before, during, and after the climate action?
How can we design a climate action so that people feel like they belong to the group, feel good about the group, and find meaning through the group's collective action? How can we communicate this to outsiders?	How can we frame our collective action so that it motivates and tells engaging stories?	Which coping strategies could help us to face the potential failure of our collective climate action?

Figure 7.1: Canvas for brainstorming and planning a specific collective climate action

Part 2

CULTIVATING RESILIENT AND EFFECTIVE COLLECTIVE CLIMATE ACTION

8

RESILIENT COLLECTIVE ACTION AND ACTIVIST BURNOUT

DOI: 10.4324/9781003558439-10

DON'T UNDERESTIMATE ACTIVIST BURNOUT

There is one consequence of collective climate action that needs to be considered, which, thankfully, the climate movement is beginning to address more: activist burnout. Some of you may be thinking that the climate crisis is so urgent – that people and nature are already suffering so much – that we don't have time to deal with our personal problems. But that is exactly the problem: activist burnout is not just a personal problem; it is also a collective problem of groups and social movements.

Activist burnout seems to be a widespread phenomenon. One study on trade union activism, for example, found that 47.5% of all activists involved in the study reported burnout symptoms.[1] It is important to actively seek ways to prevent activist burnout, as it has serious consequences for people involved in collective climate groups, as well as for the groups themselves. Activist burnout is a reason for dropping out that should not be underestimated. Dropping out is often not a deliberate decision. Rather, individuals may feel that their emotional and physical stress is forcing them to withdraw from their group.[2] For many affected by activist burnout, there seems to be a "moment of truth" when they have to make a final decision:[3] either they remain fully committed to the climate action group and push their limits thereby risking their own health or they leave the group. In the case of the first option, the activist is likely to lose patience, focus, and creativity, which negatively impacts a group and could even end up demotivating other members. In the case of the second option, the activist loses their connection to group members and what has likely become a core purpose in their life. These seem the only two paths available to individuals experiencing burnout because group structures often don't allow people to chart a third middle course.

Burnout can have cascading effects within groups and movements. If one member of a group drops out, the workload placed on other members may increase, which in turn may increase their risk of burnout.[2] In this way, burnout can breed burnout. As there is still a lot of silence around activist burnout, dropping out often comes as a surprise.[4] Activist burnout thus has a disruptive character with potentially high costs for the stability of the movement.[2,5] It is also not unlikely that those who drop out have been involved in the movement for some time – with their departure, they also take with them relevant practical knowledge, skills, and wisdom about the historical developments and positions of the group.[2,6]

As if all that were not enough, activist burnout could also affect the attractiveness of the climate movement. While collective climate action itself may be a useful coping response if a person has climate anxiety,[7] activist burnout within a group may prevent people joining collective climate action in the first place. For example, if an individual has many friends involved in a climate action group but these friends often seem depressed and close to burnout, that individual might think twice about joining the group. To the knowledge of the Author Team, this path has not been studied empirically yet.

While burnout has become a crucial issue in many work-related fields in capitalist societies, members of climate action groups seem to be particularly vulnerable to burnout due to the magnitude of the climate crisis. Within social movements, the work simply never ends and is inherently associated with numerous negative emotions.[3]

> It is indeed neither a sprint nor a marathon but a lifelong hike for which climate action groups and their members must be prepared.

People involved in social movements are the ones in our societies who are "willing to face what others are unwilling to face",[3] which, even without the workload and conflicts that come with their commitment, is already a huge burden. Thus, an enduring and effective climate movement must be a resilient movement that allows everyone involved to lead a healthy life.

 Box 8.1: Note – Trigger warning

Activist burnout is a highly sensitive issue. If you have experienced it, the following section may trigger uncomfortable emotions. With the purpose of conveying scientific knowledge, this chapter describes burnout, its symptoms, and its causes using a more rational tone and real quotes taken from people who have experienced burnout.

DEFINING ACTIVIST BURNOUT

Activist burnout is a state in which long-term stressors affect an activist's ability to remain engaged.[2,8,9] While an understanding of activist burnout is being developed and refined, it is noteworthy that burnout in general is a rather vague concept that shares some overlap with depression.[10] Cher Chen and Paul Gorski from George Mason University developed categories of activist burnout symptoms through interviews with human rights activists, some of whom also worked on environmental issues.[3] Activist burnout symptoms can be broadly classified into physical, emotional, and motivational symptoms. Let's take a deeper look at each of these symptoms.

Physical symptoms, such as being completely exhausted, were reported by activists who identified as feeling burned out. Involvement in collective action for these activists led to chronic insomnia, eating problems, migraines, headaches, and pneumonia. This critical physical state is reflected in one animal rights activist's comment: "I'm waking up in the middle of the night thinking about how I need to do this or bring this in or what time I am meeting with these parents, and that starts repeating itself".[3] Emotional symptoms reported in these interviews included chronic stress, anxiety, and depression. One activist said, "I'm feeling totally overwhelmed by the immensity of the problems we face, but I keep pushing myself. It's like an anorexic getting thin. When you're an activist, you're never working hard enough."[11] In other areas of burnout research, cynicism is another emotional symptom.[8] Symptoms that reduce a person's motivation, such as hopelessness, despair, and feelings of ineffectiveness also color burnout. One activist's comment reflects this notion: "I'm putting in all this energy and I don't see anything changing, so [you wonder], is it really working? Does it really matter that I do all this?".[3]

It is not uncommon for members of collective climate groups to experience these symptoms from time to time. However, a long-term accumulation of the three types of symptoms could indicate that a person is close to or in the midst of burnout. This person may be faced with the decision to pause their involvement for a while if they feel their group is not providing another option. This is why it is important for groups to change their cultures and for individuals to change their relationship to collective climate action, both of which are necessary for building a resilient climate movement.

Notably, quantitative research on how best to do either of these things is scarce, even though an interview study conducted by our Author Team revealed that this topic is considered very relevant within the socio-ecological movement (see the Appendix for more on quantitative and qualitative research designs).[12] The ideas presented in this chapter are mainly based on three quantitative studies, one of which was not conducted in a climate context but focused on the Israeli–Palestinian conflict[13]. The other two studies focus on people involved in movements and groups for a socio-ecological transformation and were conducted by members of our Author Team, of which one is not yet published.[14,15] Fortunately, there are already insightful interview studies on various collective action causes (animal rights, racism, humanitarian causes), a number of experience-based books, and experiences from the climate movement that have been shared with us, which are woven into this chapter.

CAUSES OF ACTIVIST BURNOUT

Interview studies have identified three broad categories of causes of activist burnout: personal and psychological factors, group norms, and problems within groups. Of course, in real life, these categories are not clear-cut but intertwined. This is why scholars often understand burnout as a mismatch between an individual and their group.[11]

For example, a person gets involved in a local fair-trade group because they are angry at global injustices related to the climate crisis. Within this group, there are many very disheartened members who talk often about the terrible state of the world. After each group meeting, this person goes home emotionally upset and feeling like they have no way of expressing their anger. This particular person-group mismatch could promote burnout. However, if anger wasn't as central a theme for this person (personal motivation) or had the person found a group that actively deals with anger (as a social norm within the group), the likelihood of burnout might have been reduced.

Although they are interrelated, a broad distinction between personal factors, group norms, and problems within groups can be helpful in gaining an overview of the causes of burnout and in identifying ways to prevent burnout through group and personal approaches.

Personal and psychological factors

While some situations can lead to burnout in almost anyone, some of us may be more susceptible to burnout. Of course, burnout often occurs when people are facing problems in their personal lives as well, such as relationship or financial issues. Additionally, previous research has shown that several personal and psychological factors are associated with activist burnout: an extensive workload, a deep sense of responsibility and anger, weak overall efficacy combined with strong participative efficacy, tensions in private and work relationships, and traumatic experiences. The tricky thing is that sometimes these factors are precisely what motivate people to get involved in collective action in the first place. Let's explore each of these personal and psychological factors.

An extensive workload

Not surprisingly, quantitative and qualitative studies have shown an increase in the likelihood of burnout the more time an individual spends involved in a climate or environment action group.[2,15] Commitment to a climate action group often comes "on top" of regular life, at the cost of personal free time. However, one of our recent studies has also shown that not only voluntary but also paid work for socio-ecological transformation is associated with burnout.[15] This suggests that people who are pursuing a career in climate- or environment-related fields may be particularly vulnerable to burnout as they are willing to invest additional time and effort for the cause.

A longitudinal study in a non-climate context also found that more collective action is generally associated with more burnout.[13] However, over time, more burnout led to less collective action, likely because people stepped away or dropped out due to activism-related stress.

A deep sense of responsibility and anger

The most commonly cited reason for activist burnout in interview studies is a deep sense of personal responsibility for the cause.[2] In the context of the climate crisis, people's sense of responsibility may be based on a deep understanding of the extent of climate change and its consequences. This responsibility may not only come from within but can also be ascribed from the outside through narratives like "the *Fridays for Future* generation will figure it out".

In one study, animal rights activists with burnout experience reported that they would say "yes" to any action that could help animals[2] – they felt strong commitment, guilt, and like they were responsible for everything. One participant remarked,

> there is a certain type of personality that is drawn ... to animal rights activism. I mean these are people who are overly empathetic, and I say 'overly' not in a bad sense ... They ... have their hearts permanently wide open and everything is flooding there: all the suffering.[2]

It is possible that feelings of guilt mask deep sadness.[4]

As this sense of responsibility is often rooted in injustice, it's not surprising that anger is associated with activist burnout.[3,13,15] Indeed, this is the most robust finding in the few quantitative studies that have been conducted. Also, research has found that young people's negative climate-related feelings, like anger, are affecting their daily lives and functioning and that they are struggling not to be overwhelmed by anger.[16,17]

Some people may also feel that they are stuck in anger, that anger seems to be constantly present. Such was the case for one respondent, who explained, "it might be anger at the system or it might be anger at individuals who I feel are part of the problem or impeding progress, kind of this background anger".[3] After all, the climate crisis is not a one-time incidental occurrence of injustice but comprises many structural injustices, which means that anger can build up again and again. Watching politicians applaud themselves for passing a climate resolution that doesn't even come close to meeting climate targets, hearing how extreme weather events have caused yet another massive crop failure in the Global South, experiencing how climate action groups are exposed to increasingly harsh repression by the police – taken together, these injustices can be too much to cope with. As participation in climate protests is closely linked to feelings of anger, people who take on the role of *rebels* in the climate movement may be particularly vulnerable to developing burnout (see Focus 2 – Strategy 4 in Chapter 5 for more on the role of *rebels*).

There is one point where current qualitative and quantitative findings don't agree. In some qualitative interviews, people attributed their deep feelings of responsibility and anger to their strong emotional attachment to the cause (for example, animal welfare). For this reason, these studies see deep emotional attachment as a source of burnout,[2] possibly driven by the perception that the cause is one's core purpose in life. However, one of the Author Team's quantitative studies found the opposite. When people are active because they feel it is their core purpose in life, it is associated with less burnout.[15] It could therefore be that the problem is not the importance people involved in the climate movement attribute to their collective climate action, but rather the constant negative emotional states of guilt and anger that cause people to experience burnout.

Weak overall efficacy combined with strong participative efficacy

A core feature of activist burnout, which is often seen as a symptom rather than a cause, is hopelessness. Both qualitative and quantitative research shows that a lack of hope, enthusiasm, and self-efficacy are associated with activist burnout.[2,14,15,18] Furthermore, one of the Author Team's recent studies found that people were more likely to experience burnout if their involvement in the movement for a socio-ecological transformation did not meet their need for competence and efficacy.[15] Interviewees report that these feelings of low efficacy arise when they are confronted with public apathy and really slow progress.[19] As transformation is not a linear process but involves setbacks and periods of stagnation, being active in the movement means having the same conversations over and over again, which can be exhausting in the long run.[20]

Such feelings of low self-efficacy are sometimes combined with a strong sense of participative efficacy, as reflected in this activist's statement:

one thing I think that doesn't always get the best audience is this kind of big picture look at [activists'] roles in addressing injustice and in making change in the sense of the smallness of our role – the idea that we can only do what we can do. With some people that's empowering and for some people it's frustrating.[18]

Members of collective action groups who experience burnout often pressure themselves to make a significant contribution to the cause.[21] They feel that their own group is highly dependent on their (extensive) involvement, and that if they were to cut back or withdraw, the group might never achieve any of its goals or even that it might cease to exist altogether. Feelings like these are demonstrative of a sense of strong participative efficacy probably having turned from being a motivator to being an obligation.

The link between participative efficacy and activist burnout was also reflected in a quantitative study on members of student sustainability initiatives.[14] It can become even more pronounced if people don't trust in the competencies of other group members, as reflected in this statement from the Master's thesis of one of our authors: "that's about what the movement could do: develop [a] more excellent cadre. So that even vain, narcissistic me doesn't feel the need to be there."[22] The initial findings in these studies suggest that burnout is likely to be associated with an unfavorable combination of weak self-efficacy (and hopelessness) and strong participative efficacy. Or in other words, "there's hardly a chance to change things, but if I'm not involved, we might as well give up now".

Tensions in private and work relationships

The more a person becomes involved in the climate movement, the more likely they are to be drawn into networks of like-minded people who encourage them to stay involved or even increase their involvement. While friendships in the movement may develop over time, there is a tendency for people involved in collective action to lose pre-existing friendships and report relationship problems with their families and partners, sometimes leading to a sense of isolation.[2,23]

Initially, meeting new, inspiring, and like-minded people could be seen as an upside for many people involved in climate action groups. However, this could become problematic if they feel that they are in jeopardy of losing the friendships built with these people if they reduce the intensity of their collective climate action. As one interviewee put it, "if your friend is like, 'hey... do you want to do extra leafletting or something with me?' then it doesn't just feel like you're saying no to the cause, you're also saying no to one of your friends".[2]

Additionally, some activists may experience criticism outside of the movement. In interviews, anti-racism activists reported feeling undervalued for their activism at their jobs, as well as feeling more vulnerable to losing their jobs as a direct result of their activism.[20] In our own experience, being involved in collective climate action is also frowned upon in some fields of academic research (even within environmental psychology).

Overall, it seems there might be a greater danger of activist burnout when climate activists find themselves in conflicts of conscience with friends or in work conflicts due to their involvement in climate action networks.

Traumatic experiences

Finally, involvement in the climate movement can be accompanied by traumatic experiences from physical threats, police violence, and other forms of repression.[2] For example, police contact during the *Ende Gelände* action depicted in Image 8.1 may well have been a traumatic activist experience, which is why organizations such as *Out of Action* have even specialized in emotional first aid during and after collective actions.

Interviews with animal rights activists in a Western country found that only a small number of activists linked their burnout to such traumatic experiences, possibly because they knew what they were getting into.[2] However, it is worth acknowledging how activists in countries with more oppressive regimes may experience trauma and consequent burnout. According to *Global Witness*, an organization exposing human rights violations, 200 land and environmental defenders were killed in 2021 by businesses, non-state actors, and governments.[24] More than half of these individuals were killed in Mexico, Columbia, and Brazil, and a quarter of these individuals were small-scale farmers. There were no documented killings in Western countries, but of course

Image 8.1: Police forces detaining an *Ende Gelände* activist as they were crossing a highway to get to the coal railway tracks they intended to block, Rhineland, Germany (2018).

Photo by Channoh Peepovicz (CC BY-NC 2.0)

other forms of repression (such as prison sentences or physical police violence) can also induce trauma.

 Box 8.2: The bottom line

The risk of activist burnout could increase when people have an extensive workload, feel very responsible and angry, generally have little hope, and perceive high participative efficacy. Traumatic experiences and excessive involvement in activist networks could also increase the likelihood of burnout.

Group norms fueling burnout

What people often don't expect when they start their engagement in collective action is to face problems with group cultures and group members.[2] As previously stated, activist burnout can stem from a mismatch between a person and their climate action group – including its group culture. Particular group cultures can lead to the impression that "we are consuming our way through people as if they were coal and oil".[25] Two types of group cultures, characterized by specific social group norms, have been associated with activist burnout. These cultures reflect previous findings on guilt and efficacy beliefs: a culture of martyrdom and a culture of competition and performance.

Culture of martyrdom

While people involved in climate action groups may already bring with them a strong sense of responsibility for the world, group cultures can reinforce this belief. One such culture, as defined by Chen and Gorski, is that of martyrdom, in which group members are made constantly aware of their guilt and even more commitment is demanded from them as a form of atonement.[3]

In one interview, an animal rights activist described a culture of martyrdom as a cause of their burnout: "I certainly think there's a sense of guilt. Anything that is not ... being an activist is considered a luxury or privilege or something that does not benefit animals or children or women or whatever the social justice issue is."[3] Here, the interviewee seems to be suggesting that their group imparts certain norms about what members of collective action groups should and should not do. A study of *Amnesty International* members found evidence of norms that activists should not focus on self-care as this would contradict their inherent selflessness.[4] Another study suggests that activists are instead urged to constantly compare themselves to the oppressed and be "super-humans [who] work unsustainable hours with hopes of creating instantaneous social change".[3]

Adding to this, groups within social movements are often faced with a lack of resources and funding.[3,19] A group with a culture of martyrdom might expose their members to financial vulnerability – jobs are often underpaid and temporary[2]. Indeed, interviewed animal rights activists described how their "supervisors treated them as though they were lucky to have jobs in the movement, as though they should be happy to have that opportunity".[2] While the amount of resources and funding available varies, groups can choose the degree to which they pass the stress of any deficiency on to their members, potentially preventing burnout. This may be easier said than done, something we at *Wandelwerk* understand well, given our own experiences with dependency on two- to three-year funding schemes, making it hard to institutionalize our structures.

Culture of competition and performance

We should not forget that climate groups are often embedded in neoliberal societies that promote cultures of competition and performance. What is more, the urgency of the climate crisis seems to reinforce these cultures within climate groups.

In interview studies, a culture of competition and performance was associated with activist burnout.[18] Some activists felt that their actions were never effective enough for their group; others experienced time pressures and did not feel appreciated for their efforts and achievements.[3,26] As one activist explained, their group culture "encourages activists' tendencies to impose unrealistic expectations on themselves, then to blame themselves when they prove incapable of meeting them".[2] As covered in Chapter 5 on efficacy beliefs, unrealistically high goals and internal attributions of failure, such as those mentioned in this quote, are powerful ways to undermine people's efficacy beliefs.

At the same time, people might compare themselves with other group members. One interviewee said,

> there is a shame to it ... [M]y friends are finding ways to push through it. They are doing [...] difficult work also, and yet I was just kind of falling apart. Makes me wonder if there is something chemically or biologically wrong with me that made me more susceptible to that type of depression – I had really severe depression – or if I just wasn't being tough enough.[2]

Everyone's limits are different. And if the social norms within a climate action group include not to speak about activist burnout, there is no way for its members to know if other members are silently suffering from similar feelings of being overworked.

A culture that encourages individuals to compare themselves with others can force people to push themselves to their limits. And a culture of competition is not limited to inter-member competition within a single group. Groups can also compete with each other; for example, for resources, for funding,[2,3,19] or for being perceived as the most successful group within a movement. Inter-group competition can have odd consequences – for example, devaluing progress if it is achieved by another group.

Ironically, bringing up individual cases of activist burnout can also be seen as a symptom of a culture of competition and performance, as problems which are actually rooted in societal structures are individualized. In other words, the focus of alleviating burnout becomes what can *you* change instead of what can our *group* change.

 Box 8.3: The bottom line

When groups promote cultures of martyrdom or competition and performance, they foster burnout among their members, as these members can become lost in feelings of guilt and the pressure to perform.

Problems within groups kindling burnout

While group norms can potentially fuel burnout, conflict between two people or between the group and an individual member can kindle it. Let's take a look at four types of problems identified in research that might lead to these intragroup conflicts: trouble with social identification, interpersonal conflicts, discrimination, and lack of interpersonal support.

Trouble with social identification

Quantitative studies have consistently found that activist burnout is associated with low social identification with the group or movement.[13,15] This may be because people feel excluded and not really valued by other group members.[15] For example, an individual might perceive fellow group members as cold, which misaligns with their need for belonging. Because of this misalignment, this individual's involvement becomes less enjoyable and their intrinsic motivation for the cause diminishes.

One study found that when people experience an emotional mismatch between themselves and the group, burnout becomes more likely.[13] For example, group members may feel that their own emotional experience of anger is not reflected by other members. This could indicate that although anger is associated with burnout, climate activists who feel anger may be less vulnerable to burnout if group members share their experience.

Interpersonal conflicts

Whether in the workplace or in the field of activism, there are factors that generally kindle burnout in groups. While occasional conflicts among group members are inevitable, how these are handled can make the difference between continuing to identify with a group and losing one's sense of belonging. Membership in climate action groups may be more flexible than professional employment. This flexibility might make it is easier for individuals to switch groups if conflicts arise with other members. Yet, interview studies have found that many activists affected by burnout reported interpersonal tensions, hostility, infighting, and bullying.[2,3,19] Some interviewed activists reported conflicts that were rooted in diverging opinions about the methods and goals of collective action (see Focus

2 – Strategy 2 in Chapter 5 for more on goals of climate action groups).[27] As one interviewee explained,

> part of [my burnout] might have been infighting, because you have to already defend and explain animal rights issues to people who are not within the larger movement. If you also have to defend the tactics or things the organization is doing within the movement, then that ends up being an issue.[2]

In other cases, the source of conflict may range from personal dislike to extensive structural discrimination.

Discrimination

Racial and sexual discrimination are recurring themes in many interview studies on collective action groups.[3] Of the 13 female activists interviewed in a study on burnout in animal rights activism, 8 said they had faced sexism in interactions with fellow group members.[2] These activists described a male group culture that they had not expected when first entering the movement: "with the movement being made up of primarily women, it's weird to feel excluded when you are in the majority and [it's weird that] men […] mostly get leadership positions".[2] Another study on burnout among racial justice activists found that all activists of color identified racism *within* their organization as a cause of their burnout.[20]

It is thus likely that already marginalized groups in our society, such as those who identify as people of color, women, people with disabilities, or LGBTQIA* or people with lower education and income are more vulnerable to experiencing activist burnout.

Lack of interpersonal support

Some interview studies have also described how cultures of martyrdom and competition prevented group members from feeling empathy for and supporting each other.[4] Imagine watching a film on climate change with your climate action group. Afterwards, you feel deeply saddened and want to talk through your concerns, but none of your group members seem to care. You leave feeling more alone in your experience than you did before meeting up with your group. In this example, sadness is responded to with apathy, and that could incite burnout. A somewhat more extreme example of this comes from activists traveling abroad to gather information on human rights abuses. After returning home, these activists reported that they felt they did not receive enough empathy for their sometimes traumatic experiences,[4] which may have led to burnout.

What is more, some activists experienced a particular lack of support when facing the first symptoms of burnout. They described receiving no mentoring on how to deal with burnout symptoms.[3] One interviewee even reported having been given the counterproductive advice to just "man up and deal with it".[3] This bears some connection to the finding that traditionally male normative concepts discourage seeking professional help for mental health problems[28].

It takes courage to raise the issue of activist burnout in cultures of martyrdom, competition and performance, which tend to silence it. When a person is then left with the idea that burnout is simply their personal problem, burnout may be just around the corner.

 Box 8.4: The bottom line

Difficulties with identification, interpersonal conflicts, discrimination, and a lack of interpersonal support within groups can kindle activist burnout among members.

BUILDING RESILIENT CLIMATE ACTION GROUPS

This overview of the causes of activist burnout raises an important question: how can we prevent burnout and build a resilient climate movement? The answer isn't a simple one. While some scholars suggest preventing burnout can only be a good thing,[3] the Author Team is a bit more hesitant with this assumption. We rather see an activist burnout dilemma.

> The activist burnout dilemma stems from the fact that some of the factors that promote burnout also promote collective climate action.

Indeed, two factors typically promoting collective climate action, anger and participative efficacy, were also found in quantitative studies to be related to activist burnout regardless of how much time a person spent in the movement.[14,15] This goes to show that certain causes of burnout also have positive aspects, making the more relevant question: how do we prevent activist burnout without killing motivation for collective climate action?

The answer to this question can lie in building a resilient movement that actively considers the relationship between members and their group and shows how the group and its members can promote resilient engagement. Resilience is when an individual, through psychological and behavioral flexibility and action, successfully manages difficulties,[29] such as burnout-promoting circumstances. Importantly, a growing body of literature suggests that resilience can be seen as a dynamic skill rather than a fixed trait that one either has or does not have.[30]

As we, the Author Team, are not aware of any research on this topic in the field of environmental psychology, we have built on our overview and compiled our own ideas and suggestions for promoting resilient collective climate action through group- and individual-focused strategies.

Box 8.5: Take action – Strategies for resilience

But first, it's your turn! If you are already active in climate action, you may already have strategies for resilience in the face of individual- or group-related issues. If you don't, or if you are not yet active in climate action, you may want to prepare some strategies.

Ask yourself:

- *What signs of over-work have I exhibited?*
- *What is my outlet for anger?*
- *How do I handle my own struggles with efficacy?*
- *What non-activist networks do I want to maintain?*

If you are part of a climate action group, also ask yourself:

- *Do we practice a culture of martyrdom and competition within our group, and if so, how can we address it?*
- *What strategies do we have for dealing with infighting?*
- *How can we sustain the health of each of our members and our group as a whole?*

Group-focused strategies for resilient collective climate action

There are a number of strategies for building resilience within climate action groups: we can give members a sense of belonging, celebrate group successes, create fun moments, let emotions shine through, develop established procedures for handling conflict, reflect on cultures of martyrdom, competition, and performance, be kind to other groups within the climate movement, assess group resources, and make healthy engagement an active subject of discussion, to name just a few. Let's delve into each of these strategies to help us gain a better understanding of how to put them into action.

Giving members a sense of belonging

Strong identification with a group may be one of the most important buffers against activist burnout.[13,15] To foster this identification, groups can actively work with the different roles and life circumstances of their members, such that, for instance, a single parent with a full-time job doesn't feel excluded.

A sense of belonging can also be rooted in mutual trust. Trust-building elements can thus be particularly valuable for groups, ranging from physical activities like "trust falls", to practicing a group structure like sociocracy, which rotates responsibilities from member to member. Indeed, one study showed that community-building strategies helped people to overcome burnout.[31] Given its importance, Chapter 2 is wholly dedicated to social identification and contains a number of ideas on how to not only promote identification but also prevent burnout.

Celebrating group successes

Social change is slow and non-linear. This makes it all the more important to be aware of small successes that support feelings of collective efficacy. One method for doing this might be holding reflection meetings after collective climate action events, in which time is specifically set aside to reflect on what went well and what was achieved, even if progress feels small.

At our regular meetings, *Wandelwerk* includes a round-of-applause item on the agenda, which gives everyone the opportunity to acknowledge things the group and its members have achieved or put effort into since the last meeting.

Creating fun moments

Ensuring members enjoy being part of the climate movement is an important part of keeping it healthy. Within groups, it could be valuable to see living with joy, wellbeing, and happiness as an act of self-liberation and resistance against an otherwise dour society. Having fun is obviously accompanied by many positive feelings, including a sense of vitality and a greater overall satisfaction with life[15] – both of which we probably all want for our fellow group members. Therefore, anything that promotes fun, such as creating room for humor,[31] playing games, organizing meetings in a playful online space, or designing a stress-relieving check-in and check-out process, may help create a regenerative culture and prevent burnout.

Given that we tend to perceive activities as fun when they meet our needs, check out the various strategies for need fulfillment scattered throughout the other chapters. A good place to start is the section on creating groups that meet people's needs in Focus 2 of Chapter 2.

Letting emotions shine through

When we perceive that our emotions are similar to those of other group members, we are less likely to experience burnout.[13] One method for helping members realize how fellow members are feeling is to hold group meetings in which emotions can be shared and similarities discovered. Experiences like these meetings might help people feel less alone in their emotions. At the same time, learning about the struggles of fellow members may pave the way for emotional group support.[31]

Some climate action groups utilize so-called heart-sharing sessions, in which each individual is given a few minutes in which only they can speak, while everyone else listens carefully. Similarly, some methods from environmental activist and deep ecologist Joanna Macy's book, *Active Hope*, involve one person describing what they're grateful for or what's causing them to feel despair while another person listens attentively and without commenting.[32] Macy's books and courses offer many methods for getting in touch with our emotions in the context of the climate crisis. Exercises like these may help align members' emotions and create a basis for a supportive group culture.

Developing established procedures for handling conflict

Because disagreements over methods and goals, as well as other intragroup conflicts, are such major causes of activist burnout, it is crucial that groups develop established procedures for revealing conflict and dealing with it. Points of contention have a way of festering just under the surface. To alleviate this, it's important to create situations in which conflict and criticism can be appropriately expressed. And once issues are brought to light, groups need to consider whether they can deal with them internally or whether they would benefit from the help of an external mediator.

As various forms of discrimination, such as racism and sexism, are likely to be present more or less subtly in every climate action group, groups should actively reflect on these highly sensitive issues. Group members could, for example, attend seminars focused on dealing with discrimination.

Reflecting on group cultures of martyrdom, competition, and performance

Some of the group cultures that promote activist burnout are a fairly direct reflection of contemporary society. Especially in the context of the climate crisis, a lot of effort has been put into making people believe that it is their individual responsibility (and not the responsibility of governments or corporations) to tackle the climate crisis. Members of climate action groups are likely to already feel sufficient responsibility and do not need to be further pushed by a culture of martyrdom.

The same applies to a culture of competition. Most activist groups and their members already feel such a degree of urgency[3] that no additional pressure through competition is needed. However, climate action groups should not blame themselves or their members for practicing these cultures as they are prevalent in society at large, with competition, domination, acceleration, and discrimination shaping people from an early age. Rather, it may be useful to reflect on what societal trends the group wants to avoid replicating.

Cultures of competition can also have a negative effect on the movement overall. Disparaging talk towards other groups pursuing essentially the same goals, for example, unnecessarily strains capacities within the climate movement. All groups can draw strength from focusing on supporting each other and showing solidarity. Even if one group doesn't fully agree with the position or actions of another group, it may be worth focusing on what unites the two rather than what divides them.

Assessing group resources

With big tasks ahead of us, we sometimes lose sight of the practical demands and requirements. This is why it's important to critically examine a group's resources (time, people power, financing) before deciding to do an action.[3] Sometimes, one less event can mean more healthy members and thus a resilient group in the long run.

Table 8.1: Cultural change for resilient climate action groups

Current culture		Prospective culture
culture of martyrdom	→	culture of belonging
culture of competition	→	culture of appreciation and celebration
culture of contained emotions	→	culture of sharing emotions and support
culture of infighting	→	culture of organized group conflict
culture of silencing burnout	→	culture of active dialogue about burnout

Making healthy engagement an active subject of discussion

Even if you are not currently experiencing burnout, someone else in your group might be. This is why the issue of activist burnout should be actively addressed in every climate action group. Fortunately, some organizations are already working on burnout prevention programs that individuals and groups can participate in. Prevention can also cover changes in group culture (see Table 8.1 for proposed changes to certain group cultures for resilient climate action groups).

 Box 8.6: The bottom line

Previous research on activist burnout offers valuable insights for how to design resilient groups. Ideally, these groups should create a basis of identification for their members, celebrate successes, plan fun experiences, facilitate emotional exchange, establish procedures for handling conflict, reflect on their group cultures, cultivate solidarity with other climate groups, regularly assess their resources, and look for ways to stay healthy together.

Individual-focused strategies for resilient collective climate action

> "Caring for myself is not self-indulgence, it is self-preservation, and that is an act of political warfare."[33]

These words by Black civil rights and feminist activist Audre Lorde raise a relevant point: engaging in burnout prevention can be a political act, which may help some people accept it as necessary for a resilient climate movement.

Individuals involved in climate action groups have at least some control over their physical, emotional, and motivation-related state of burnout, the causes of which can potentially be alleviated by employing various strategies. We can make time for recreation, reflect on the right balance of motivation, blow off activism steam, increase feelings of overall efficacy, say no, disseminate tasks, trust others, maintain non-activist networks, scale down, switch to another type of action, plan to leave, or consider therapy. Let's take a look at each of these strategies.

Making time for recreation

For many members of the climate movement, taking time for leisure seems like taking time away from collective climate action. For these individuals, the climate crisis seems too urgent to "waste" even a second. If this aligns with how you feel, please consider taking a look at the section on the consequences of activist burnout described at the beginning of this chapter.

Since something as important as collective action tends to absorb any free time, it can be useful to plan regular time for recreation and positive experiences.[31] Take a moment to think about what you would ideally fill this time with. For one person, it might be letting go of pent-up frustration on the basketball court. For another, it might be painting in a peaceful studio. For another still, it might be meditating in a cozy space.[34] While connecting people's passions with collective climate action might be a good way to motivate them to participate in collective climate action, to prevent burnout it may be useful to fill one's free time with something completely unrelated to the cause. Creating mental distance from our activist passion may be key.

Another way to spend one's free time could be in nature.[11] Many people involved in the climate movement are driven by a deep connection to nature. Experiencing the positive sides of this bond could also be crucial to giving new meaning to one's climate action.[35]

Reflecting on the right balance of motivation

One interpretation of our overview of activist burnout is that there may be a healthy and resilient amount of anger, guilt, and efficacy for each person that keeps them engaged while protecting them from burnout symptoms. This healthy level would strike a balance between a person's motivating factors for collective climate action.

An example of imbalance would be a person whose motivation for collective action is based entirely on feeling indispensable to their climate action group, which could lead them down the path of burnout; an example of balance would be a person whose motivation is based on a blend of anger, guilt, participatory efficacy, and identification with the group and the movement – a mix that might make them better equipped to avoid or handle burnout.

It may be helpful for members of collective climate groups to reflect on their motivation and, in so doing, find a balance that feels healthy. Here are some questions to ask yourself to get started in this reflection process:

- *How much anger and injustice do I feel, and can I live a happy life with this amount in the long run?*
- *What role does guilt play in my climate actions? Do I do some things only because I feel obligated?*

- *Do I think my commitment is integral to the survival of my climate action group?*
- *Given that it's not bad to derive some self-esteem from our roles in collective action, what do I see as a good amount of participative efficacy for me in my group?*

Blowing off activism steam

While collective climate action has the potential to generate anger, it may also be a way of blowing off steam. Recent research found that people who felt or could express climate-related anger reported lower anxiety, depression, stress, and guilt.[36,37] This fits a personal account from one of our workshops, in which a participant explained that although she did not believe in the effectiveness of protests, she still continually joined them as a way to shout out her anger.

On the other hand, actively suppressing anger could actually wind up exacerbating it. This is like trying to hold a basketball down under water – the deeper you try to push it down, the more difficult it is to control. If you don't exert constant effort to keep the ball submerged, it's going to slip from beneath your hands and shoot up to the surface. Instead of expending the effort to keep the basketball of anger forced down at all times, we could try instead to let it rise to the surface. In the case of climate-related anger, actively dealing with it might transform it into collective climate action.

Engaging in normative protests, as opposed to non-normative actions, could be especially suitable for blowing off steam, as the likelihood of being confronted with new injustices such as police violence is smaller.

Increasing feelings of overall efficacy

Chapter 5 presents a number of strategies for promoting feelings of efficacy. Of those, appreciative feedback, positive visions, and enthusiastic environments will most likely prevent burnout – as long as they don't lead to an individual feeling too indispensable and they also highlight the collective efficacy of the whole group[14].

Saying no, disseminating tasks, trusting others

To reduce one's own responsibility in a group, it can be useful to learn to say no to things. In the same vein, Christina Maslach and Michael Leiter from the universities of California and Acadia describe an example of an environmental activist who worked on being able to disseminate tasks he once felt he had to do alone.[11] At first, he was reluctant to do this because the cause was so important to him. When he finally unloaded some tasks onto others, he started developing trust in his fellow group members. Our own research as well has shown that individuals trusting in the competencies of fellow group members report experiencing less burnout.[14]

Maintaining non-activist networks

Research suggests that movement members should not lose sight of their non-activist networks.[31] It may be easier to feel comfortable around like-minded people who share the same strong urge to make the world a better place, but non-activist friends and family can be good companions in times of much-needed rest. Non-activist networks can also provide support when a movement member needs to reduce their involvement due to group-related stress.

Of note, it may be useful to practice *non-violent communication*[38] when attempting to maintain positive relationships with those outside our activist networks, particularly when our more deep-set opinions diverge. For more on how to challenge but not threaten another's moral self-image, see Focus 2 – Strategy 2 in Chapter 3.

Scaling down, switching to a different type of action, or planning to leave

Activist burnout might occur when a person gets stuck in a particular form of collective climate action or in a role within the climate movement that does not suit their physical, emotional, or motivational state. There are countless stories of activists who were once on the front lines of protests, but after experiencing burnout, starting a family, or intense self-reflection moved into roles in the background. Switching to a new type of action or to a new role could create a better fit between a person's current life circumstances and their group. Of course, these new elements are not set in stone – later reflections or changes in life circumstances could lead to further changes in actions and roles down the road.

Planning to leave certain activities or groups may be another strategy to prevent burnout. Before doing so though, it might be appropriate to pass on any knowledge or responsibilities in order to avoid negative consequences for the climate action group. Retreating from certain activities or groups can still be considered a functional strategy within the framework of an overall resilient movement.

Considering therapy

Researchers also suggest therapy as an approach for coping with activist burnout.[3] As symptoms of burnout and depression can overlap,[10] therapy may be a valid approach to confront burnout. However, people involved in collective action may need to search for a while to find a therapist capable of not individualizing their problems but rather taking a systemic perspective.[12]

Those who have experienced traumatic situations, in particular, may benefit from speaking with a professional.

Where individual-focused actions become group-focused actions

Many of these individual-focused strategies for resilient collective climate action can also be applied to group settings, and they may even be more effective if discussed with fellow group members. For example, groups can try to distribute tasks among

Image 8.2. Participants of an *Ende Gelände* action relaxing in hammocks, Germany (2018).

Photo by Tim Wagner (CC BY-SA 2.0)

members in a needs-based way. They can ensure that members have time for recreation (as depicted in Image 8.2). They can organize meetings in which members can reflect on their motivation and their support network and learn to say no. And groups can try to create environments that foster feelings of overall efficacy.

 Box 8.7: The bottom line

In addition to group strategies, people involved in collective climate action can also engage in burnout prevention at the individual level. This could be described as a political act, which may help some people accept it as necessary for a resilient climate movement. The likelihood of activist burnout could be reduced through regular leisure time, a suitable balance of motivation types, moments to let off steam, strategies to promote self-efficacy, learning to say no, task distribution, trust in the group, non-activist networks, a reduction and change in engagement, or therapy.

GROUP EXERCISE FOR ASSESSING ACTIVIST BURNOUT

Based on the quick burnout assessment by Maslach and Leiter,[11] the Author Team has developed a group exercise for assessing activist burnout. This exercise, which involves the use of a questionnaire you can download from the *Wandelwerk* website,[39] is broken down into three steps designed to help groups develop resilience by identifying their strengths as well as potential problems that may cause burnout.

Step 1: Fill out the questionnaire
On your own and without sharing your responses with fellow group members, fill out the questionnaire shown in Table 8.2 (or something similar you designed

Table 8.2: *Questionnaire for assessing activist burnout based on Maslach and Leiter*[11]

	With respect to this point, my involvement in the climate group is …		
	just right	mismatch	major mismatch
Workload The amount of work for the group in a week			
Leisure time Free time for nature Free time for hobbies and exercise Free time for contacts outside the group			
Guilt and anger Moments in which I feel guilty Moments in which I feel angry about climate injustices Opportunities to let off steam			
Hopefulness Moments in which I feel hopeful Moments of (small) success			
Participative efficacy Responsibility that I have in my group (Time) pressure that I perceive in my group			
Support networks Emotional support I receive in my group Practical support I receive in my group (e.g., task support)			
Identity My own closeness to the group Appreciation I receive Possibilities to share my emotions with others			
Conflicts The amount of conflicts Times I feel angry towards other group members The group's way of dealing with conflicts Discrimination that I perceive in the group			

on your own). Evaluate each sub-category (such as free time for nature, hobbies, contacts) separately and indicate your answers on the same line height. As a group, you should decide beforehand whether you want to have the questionnaires remain anonymous.

How you fill out this questionnaire depends more on your unique needs and perceptions than actual facts. In each row, indicate whether your participation in a climate action group is just right, a mismatch, or a major mismatch. For example, checking "just right" in the "free time in nature" row might indicate that your involvement in the climate action group leaves you enough time to spend in nature; checking "mismatch" might indicate that you haven't had time to get outside for the past week and it is starting to annoy you, while "major mismatch" might mean that you haven't had time in nature in months due to your group obligations, and you're strongly suffering because of it.

Step 2: Interpret the questionnaires
Compile and shuffle all the completed questionnaires. Redistribute the questionnaires. In small groups, examine and interpret the responses and develop strategies for what needs to change to prevent burnout. It's important to note when interpreting the results that a few mismatches are typical.[11] However, if several group members checked a mismatch or if there are major mismatches for a few members, these may be relevant to look at for burnout prevention.

Step 3: Rejoin into one large group
Share the ideas from your small group work and come up with plans for how to implement them. This step should provide individuals with the opportunity to reflect on how well or not well they align with the group and fellow members. And it should provide groups with the opportunity to reflect on how their structure aids or hinders fulfilling the needs of their members.

References

1. Nandram, S. S. & Klandermans, B. Stress experienced by active members of trade unions. *J. Organ. Behav.* 14, 415–431 (1993). https://doi.org/10.1002/job.4030140504
2. Gorski, P., Lopresti-Goodman, S. & Rising, D. "Nobody's paying me to cry": The causes of activist burnout in United States animal rights activists. *Soc. Mov. Stud.* 18, 364–380 (2019). https://doi.org/10.1080/14742837.2018.1561260
3. Chen, C. W. & Gorski, P. C. Burnout in social justice and human rights activists: Symptoms, causes and implications. *J. Hum. Rights Pract.* 7, 366–390 (2015). https://doi.org/10.1093/jhuman/huv011
4. Rodgers, K. 'Anger is why we're all here': Mobilizing and managing emotions in a professional activist organization. *Soc. Mov. Stud.* 9, 273–291 (2010). https://doi.org/10.1080/14742837.2010.493660
5. Pigni, A. *The Idealist's Survival Kit: 75 Simple Ways to Avoid Burnout.* (Parallax Press, 2016).
6. Plyler, J. How to keep on keeping on: Sustaining ourselves in community organizing and social justice struggles. *Upping Anti.* https://uppingtheanti.org/journal/article/03-how-to-keep-on-keeping-on/ (2009).

7. Schwartz, S. E. O. *et al.* Climate change anxiety and mental health: Environmental activism as buffer. *Curr. Psychol.* 42, 16708–16721 (2023). https://doi.org/10.1007/s12 144-022-02735-6

8. Maslach, C. & Gomes, M. E. Overcoming Burnout. In *Working for Peace: A Handbook of Practical Psychology and Other Tools* (ed. MacNair, R. M.) 43–49 (Impact Publishers/ New Harbinger Publications, 2006).

9. Freudenberger, H. J. Staff burn-out. *J. Soc. Issues* 30, 159–165 (1974). https://doi.org/ 10.1111/j.1540-4560.1974.tb00706.x

10. Bianchi, R., Schonfeld, I. S. & Laurent, E. Burnout–depression overlap: A review. *Clin. Psychol. Rev.* 36, 28–41 (2015). https://doi.org/10.1016/j.cpr.2015.01.004

11. Maslach, C. & Leiter, M. P. Reversing burnout: How to rekindle your passion for your work. *IEEE Eng. Manag. Rev.* 38, 91–96 (2010). https://doi.org/10.1109/ EMR.2010.5645760

12. Hamann, K. R. S. *et al.* How can psychological research support movements for socio-ecological change? A qualitative study on psychological challenges and questions of activists. *Global Environmental Psychology.* https://psycharchives.org/en/item/ 2723d856-0b7b-459c-a79f-f51c7ea529e0 (2024).

13. Vandermeulen, D., Hasan Aslih, S., Shuman, E. & Halperin, E. Protected by the emotions of the group: Perceived emotional fit and disadvantaged group members' activist burnout. *Pers. Soc. Psychol. Bull.* 014616722210928 (2022). https://doi.org/ 10.1177/01461672221092853

14. Hamann, K. R. S., Holz, J. R. & Reese, G. Coaching for a sustainability transition: Empowering student-led sustainability initiatives by developing skills, group identification, and efficacy beliefs. *Front. Psychol.* 12, 623972 (2021). https://doi.org/ 10.3389/fpsyg.2021.623972

15. Hamann, K. R. S., von Agris, A.-S. & Markus, L. Investigating the predictors of collective action intensity and health. https://osf.io/c7vsn/ (2023).

16. Hickman, C. *et al.* Climate anxiety in children and young people and their beliefs about government responses to climate change: A global survey. *Lancet Planet. Health* 5, e863–e873 (2021). https://doi.org/10.1016/S2542-5196(21)00278-3

17. Marczak, M., Winkowska, M., Chaton-Østlie, K., Morote Rios, R., & Klöckner, C. A. "When I say I'm depressed, it's like anger." An exploration of the emotional landscape of climate change concern in Norway and its psychological, social and political implications. *Emotion, Space and Society,* 46, 100939 (2023). https://doi.org/10.21203/ rs.3.rs-224032/v2

18. Gorski, P. C. & Chen, C. "Frayed all over:" The causes and consequences of activist burnout among social justice education activists. *Educ. Stud.* 51, 385–405 (2015). https://doi.org/10.1080/00131946.2015.1075989

19. Gomes, M. E. The rewards and stresses of social change: A qualitative study of peace activists. *J. Humanist. Psychol.* 32, 138–146 (1992). https://doi.org/10.1177/002216789 2324008

20. Gorski, P. C. Fighting racism, battling burnout: Causes of activist burnout in US racial justice activists. *Ethn. Racial Stud.* 42, 667–687 (2019). https://doi.org/10.1080/01419 870.2018.1439981

21. Pines, A. M. Burnout: An Existential Perspective. In *Professional Burnout: Recent Developments in Theory and Research* (eds. Schaufeli, W. B., Maslach, C. & Marek, T.) 33–51 (Taylor & Francis, 1993).

22. Junge, E. *The Wastefulness of the Environmental Movement – An Investigation into Sustainable Activism.* (Lund University, 2018).

23. Vestergren, S., Drury, J. & Chiriac, E. H. The biographical consequences of protest and activism: A systematic review and a new typology. *Soc. Mov. Stud.* 16, 203–221 (2017). https://doi.org/10.1080/14742837.2016.1252665

24. Global Witness. Decade of defiance. https://www.globalwitness.org/en/campaigns/environmental-activists/decade-defiance/ (2022).

25. Haaheim, J. Dr. King's question to today's social movements. *Justin Haaheim* https://justinh.org/2013/08/29/darkness-cannot-drive-out-darkness/ (2013).

26. Klandermans, B. Disengaging from movements. In *The Social Movements Reader: Cases and Concepts* (eds. Goodwin, J. & Jasper, J.) 128–140 (Blackwell, 2003).

27. Hopgood, S. Keepers of the flame: Understanding amnesty international. *Perspect. Polit.* 5, 213–215 (2007). https://doi.org/10.1017/S1537592707070703

28. Seidler, Z. E., Dawes, A. J., Rice, S. M., Oliffe, J. L. & Dhillon, H. M. The role of masculinity in men's help-seeking for depression: A systematic review. *Clin. Psychol. Rev.* 49, 106–118 (2016). https://doi.org/10.1016/j.cpr.2016.09.002

29. APA. APA Dictionary of Psychology: Resilience. https://dictionary.apa.org/resilience.

30. Leys, C. *et al.* Perspectives on resilience: Personality trait or skill? *Eur. J. Trauma Dissociation* 4, 100074 (2020). https://doi.org/10.1016/j.ejtd.2018.07.002

31. Nepstad, S. E. Persistent resistance: Commitment and community in the plowshares movement. *Soc. Probl.* 51, 43–60 (2004). https://doi.org/10.1525/sp.2004.51.1.43

32. Macy, J. & Johnstone, C. *Active Hope: How to Face the Mess We're in Without Going Crazy.* (New World Library, 2012).

33. Lorde, A. *A Burst of Light.* (Firebrand Books, Ithaca, 1988).

34. Gorski, P. C. Relieving burnout and the "Martyr Syndrome" among social justice education activists: The implications and effects of mindfulness. *Urban Rev.* 47, 696–716 (2015). https://doi.org/10.1007/s11256-015-0330-0

35. Westoby, R., Clissold, R. & McNamara, K. E. Turning to nature to process the emotional toll of nature's destruction. *Sustainability* 14, 7948 (2022). https://doi.org/10.3390/su14137948

36. Stanley, S. K., Hogg, T. L., Leviston, Z. & Walker, I. From anger to action: Differential impacts of eco-anxiety, eco-depression, and eco-anger on climate action and well-being. *J. Clim. Change Health* 1, 100003 (2021). https://doi.org/10.1016/j.joclim.2021.100003

37. Rothgerber, H. *et al.* Motivated moral outrage among meat-eaters. *Soc. Psychol. Personal. Sci.* 13, 916–926 (2022). https://doi.org/10.1177/19485506211041536

38. Rosenberg, M. B. & Chopra, D. *Nonviolent Communication: A Language of Life.* (PuddleDancer Press, 2015).

39. Wandelwerk e.V. Wandelwerk Umweltpsychologie: Wir bringen Psychologie in den Umweltschutz. https://www.wandel-werk.org/

9

SOCIO-ECOLOGICAL TRANSFORMATION

DOI: 10.4324/9781003558439-11

LOOKING AT THE BIGGER PICTURE

Decisive collective action in favor of climate protection can be the basis for major social change towards the world we want. In the last few years, the climate movement has experienced massive growth, and with this growth, more significant change through climate action than previously thought possible. Now, countless groups, collectives, organizations, and individuals have begun pondering how all the many forms of collective climate action could strategically combine for a greater shift.

For climate action to be successful, it is crucial that the individuals involved have an understanding of theories as well as multiple practices of social change. This understanding starts with raising questions: How does society change? What successful political changes made by collective actors have occurred in the past – and what can we learn from them to succeed with present and future actions? How can different approaches to collective climate action strengthen and complement each other and amplify the impacts of the transition? Referring back to the hiking guide mentioned in the Preface, theories of change are like hiking maps: they can help us orient ourselves and find direction for short- and long-term goals, nourish our motivation to keep going, and support us in prioritizing where to invest our energies most effectively.

While psychology is an appropriate science for understanding human thoughts, feelings, and behavior, it does not cover a wider understanding of the society we live in and how it could change. This is why this section moves beyond psychology and into other scientific and practical fields that try to understand the patterns of socio-ecological change.

Acknowledging that we as the Author Team are not experts in transition research, in this section, we present and summarize approaches that we ourselves find useful in looking at the bigger picture. These approaches include Geels and Schot's Multi-Level Perspective, the Social Tipping Point Concept, Wright's Three Strategies of Transformation, and Moyer's Movement Action Plan.[1-4] This selection includes approaches to technological transformation that are widespread in present-day policy and approaches to social transformation that are currently under-researched[5] but often discussed within movements. Alongside these approaches, there are also ideas from two German climate action groups, the *I.L.A. Collective* and *ausgeCO2hlt*. These groups apply scientific work towards socio-ecological change. We have also incorporated our own insights based on many years of campaigning and organizing within the German climate justice movement.

 ## Box 9.1: Note – Diversity of approaches to socio-ecological change

Before we dive in, it's important to acknowledge that the selected approaches come from White, Western authors. In order to grasp the true complexity of social change, it is necessary to include diverse perspectives. We believe there is strategic value in such diversity for enhancing the climate movement's capacity to promote fundamental change. Therefore, this chapter must be seen as a preliminary understanding of socio-ecological transformation and how fundamental change can be achieved on a larger scale.

DEFINING SOCIO-ECOLOGICAL TRANSFORMATION

In 2011, the German Advisory Council on Global Change put forward an understanding of a "great socio-ecological transformation" as a deep societal transformation directed at decarbonization and socio-ecological justice.[6,7] Building on research and practical experiences, this concept has been extended by the *I.L.A. Collective* to include the aim of achieving a democratic society that questions current social power relations.[8] Based on these ideas,

> socio-ecological transformations are large-scale changes aimed at promoting a socially just, ecologically stable, and democratic society.

From the perspective of transition researcher Maja Göpel, now an honorary professor at Leuphana University of Lüneburg, and her colleague Moritz Remig, socio-ecological transformation "does not consist of one big transformation [...]. [Instead,] many small sequential and parallel transformation processes in different subsystems [lead] to a change in societal development or system dynamics."[9] However, it is the sum of these many small steps that gives the transformation a direction.[8]

These processes occur and interact at many levels, from changes in individual beliefs and behaviors, shared values and norms, and relationships with others and the natural world, to fundamental shifts in the way politics, society, and the economy are structured.[2,10] In this understanding, these many socio-ecological transitions can be rapid and abrupt, sometimes invisible, in different places, driven by different actors and motivations, and can bring about fundamental change. The more we understand the interplay of different societal actors and subsystems, the better we can actively and strategically shape change instead of just reacting to it – making change happen by design instead of waiting for disaster to strike and force change upon us.

Importantly, these multiple parallel and interacting transitions are needed as an adequate response to socio-ecological problems that also arise from the interaction of many subsystems, such as politics, economics, civil society, media, and other fields linked to numerous social narratives, ways of thinking, and behaviors[11]. As Göpel points out, there is no single flaw in one of these subsystems that we can repair through isolated analysis and focusing on only one subsystem.[9] Taking this systemic perspective can help decide where to intervene.[2] The theories of change presented in this chapter build on the idea of different subsystems and actors as part of larger transformations.

THE MULTI-LEVEL PERSPECTIVE

One of today's most discussed approaches in transition research is the Multi-Level Perspective by Frank Geels and Johan Schot from Eindhoven University of Technology[1]. Drawing on institutional theory, evolutionary economics, and the sociology of technology, the Multi-Level Perspective was originally developed to explain the spread of technological innovation in Western capitalist democracies.[1] It is neither an overarching theory that attempts to explain everything nor a perspective based solely on data. Rather, it is a middle-range theory that can be applied to the development of sustainability in major systems (transportation, food, energy) and subsystems (transportation mode – train, bus, bike; transportation distance – long-distance, intercity, local).

Three levels of change

This approach is called the Multi-Level Perspective because it understands transitions as emerging from the interplay of three heuristic and analytical levels of change: *landscape, regime,* and *niche.*[12] Let's take a look at these three levels of change.

The landscape level

The first of these levels, the *landscape* level, is the largest and most enduring level, as can be seen in Figure 9.1. It consists of large-scale economic changes, political developments, and deep cultural patterns.[1] Examples include megatrends such as climate change, demographic trends such as ageing societies, and culturally held narratives such as "the free-market self-regulates".

When changes do occur at the *landscape* level, they tend to take the form of very slow developments that take decades to unfold and affect all other levels. However, there is also the possibility of rapid external shocks such as war.[12] The other levels, in turn, cannot effect changes at the *landscape* level in the short term. However, they can put pressure on developments on the *landscape level*; for example, through climate protests, mobilization of public opinion, and climate regulations.

The regime level

The second of these levels, the *regime* level, is the basis for stability in socio-technical systems.[12] It includes co-evolving and coordinating regimes such as politics, industry, technology, culture, and science. For example, an actor at this level may be a fossil fuel company with powerful influence on policy or a research group presenting strongly evidence-based climate science.

Regimes are stabilized by path dependencies, their social norms and values, shared understanding of regulations and laws, and cognitive routines such as beliefs, problem definitions, and general guidelines.[1] For example, *regimes* promote the preservation of already institutionalized democratic principles but also prevent people from imagining alternatives to our current fossil-based society.

Though quite a stable level, conditions are less enduring at this level than at a *landscape* level, as rules can be subject to debate and internal conflict.[12]

The niche level

The third of these levels, the *niche* level, comprises innovations and radical novelties that are smaller and less stable than *regimes* and emerge in spaces that are protected from mainstream market selection.[1] They are often developed by outsiders or newcomers.[13] At the *niche* level, diverse, small groups and innovations interact and aim to deviate from, add to, or replace the existing *regime*.

Niches can be economic innovations that have been developed in research and development laboratories, such as new ways of capturing wind energy. *Niches* can also be social innovations, in which sustainable forms of living and working together, as well as values such as solidarity and community, are already being lived out on a small scale. Examples of this include repair cafés, solidarity-based ecological farming models, swap stores, and neighborhood gardens. The seeds of innovation must be discovered, brought into contact with one another, and further developed for widespread use. In this way, they can grow out of their *niche* level and ultimately be accepted by a broader public.

The theory of change for the Multi-Level Perspective

According to the Multi-Level Perspective, transformation processes happen non-linearly through the interaction of the three levels of change across time and space.[13] Geels and Schot assume that transformation cannot be planned and coordinated from the onset as it depends on a co-occurrence and complex interaction of three processes, as shown in Figure 9.1.[1]

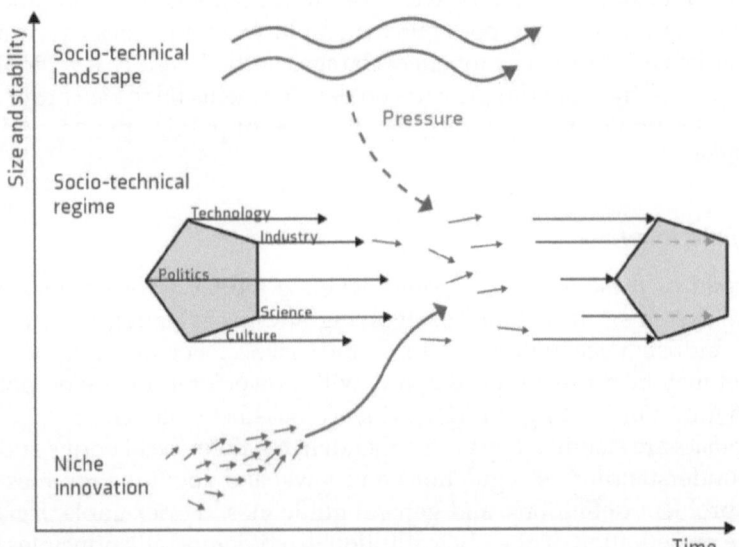

Figure 9.1: Adapted illustration of Geels and Schot's Multi-Level Perspective[1]

The first of these processes comes when *landscape* changes put pressure on a *regime*. This could be through megatrends at the *landscape* level, such as growing numbers of extreme weather events or large-scale protests and shifts in public opinion on the climate crisis. The second comes when *regimes* are destabilized, when they experience tensions and cracks. Under these circumstances, windows of opportunity emerge in which *niche* innovations such as renewable energy or new social practices can take root in *regimes*. Transformation then ultimately occurs in the form of the third process, when *niches* are ready to be included into *regimes*; for example, when wind technologies reached a certain degree of efficiency to become a part of the energy system. *Niches* can become large in scale and can ultimately replace an entire *regime*. There is also the possibility that *niches* will replace *sub-regimes*, which may then lead to long-term, sweeping transformation.[1]

Next to these three processes typically associated with the Multi-Level Perspective, Geels and Schot also describe other pathways that are likely if the aforementioned processes do not co-occur.[1] One such pathway occurs when there is simply no *landscape* pressure on a *regime*, resulting in the *regime* replicating and, if at all, changing slowly in predictable directions within the existing rules. With this pathway, it is crucial to keep in mind the manifold antagonistic and stabilizing climate-related *landscape* trends[12] – once certain structures are in place, they remain in place for a long time, even when their disadvantages begin outweighing their use. Such structures stand in the way of new, sustainable structures, through both their high costs and the space they occupy. The 1960s model of the car-friendly city, for example, continues to shape the infrastructure of many cities today. It is sustained throughout the world simply because it is already there and change would be immediately costly, even though there are many arguments against it, such as health, community, or climate reasons (that come with their own costs). This tendency of a system to remain in its current state and thus neglect possible alternatives is also known as *path dependency*.

Another pathway occurs when *landscape* pressures are moderate but not disruptive. Under these conditions, *regimes* may seek to adapt to and include initial demonstrations of *niche* innovations by entrepreneurs or activists. Here, according to Geels and Schot, "new regimes grow out of old regimes".[1] Within the climate crisis, these may be extreme weather events in countries that are not yet severely affected by climate change. It may be possible to manage each individual weather event so that there is no pressure (yet) to change the entire *regime*, but efforts increase to adapt to *landscape* changes one step at a time.

Yet another pathway occurs when sudden and large *landscape* pressures occur but *niche* innovations are not yet ready.[1] From an optimistic perspective, this window of opportunity may encourage the rapid development of multiple *niches*, one of which could be incorporated into the *regime*. However, the development of *niche* innovations is not a given. According to the Multi-Level Perspective, *niches* need to articulate visions, build social networks, and engage in learning activity in order to develop successfully.[12]

An illustrative example of a *niche* comes from an analysis of the UK transition town movement by Gill Seyfang and Alex Haxeltine from the University of East Anglia.[5] Seyfang and Haxeltine saw transition towns as *niche* innovations targeting multiple *regimes* (energy, transportation, food, housing) and emerging in response to peak oil as a *landscape* pressure. They argue that transition towns have successfully developed visions but need to develop intermediate steps that connect their small local steps to their larger vision. While the transition town movement did indeed build strong networks, these networks were largely only within the socio-ecological movement and local governments and lacked networks with other *regime* actors. Furthermore, the transition town movement put a strong emphasis on learning and training but needed to go beyond a "knowledge leads to action" approach (see Chapter 6 for more on action first, motivation second). In this way, *niche* innovation research may help in understanding the effectiveness of specific climate action groups.

Although not always easy to see, groups with agency can be found at all levels of the Multi-Level Perspective; for example, organizations, industries, policymakers, engineers, researchers, and individuals in their roles as consumers and citizens.[12] To highlight the role of individuals, Göpel has extended the Multi-Level Perspective to include a level of individuals that clarifies how people make up institutions at the *niche* and *regime* levels.[2] The perspective offered by this level makes it easier to link collective climate action with systemic socio-ecological change.[7] Community members could try out new socio-ecological ways of life that constitute *niche* innovation; citizens could get involved in politics and, together with people working in the current *regime* institutions, try to integrate new *niches* into *regime* rules and laws; climate protestors could put pressure on the *regime* by changing public opinion at the *landscape* level; sustainability educators could question dominant narratives at the *landscape* level.

 Box 9.2: Food for thought – Why is a socio-ecological transformation so difficult

The Multi-Level Perspective also offers some ideas as to why socio-ecological transformations may be more difficult than the more commonly analyzed technological transformations:

- Since socio-ecological *niches* develop in response to *regime* inaction, their inclusion into *regimes* is inherently less likely.[5]
- Compared to economic innovations, socio-ecological transformations pursue goals that are more dependent on support from public authorities and civil society, such as social justice.[12]
- Socio-ecological innovations do not always have the economic benefits that market innovations have, so their implementation is linked to changes at the economic *regime* level – for example, in taxes or regulations.
- Socio-ecological transformations require changes in *regimes* that are stabilized by various systems and dominated by key players. This can be seen, for example, in the *regimes* within the energy and transportation sectors.

These factors need to be discussed, as they illustrate precisely how significant socio-ecological transformation requires many smaller transformations at each of the three levels of change.

Box 9.3: The bottom line

The Multi-Level Perspective describes three levels that are crucial for analyzing socio-ecological transformations: *landscape, regime,* and *niche*. Transformation occurs when *landscape* changes put pressure on and destabilize *regimes*. This opens a window of opportunity for prepared *niche* innovations to replace or gradually influence *regimes*.

THE SOCIAL TIPPING POINT CONCEPT

Another stream of research currently underway is investigating the emergence of social tipping points. The concept of tipping points is commonly used in work on ecological planetary boundaries, which has aimed to define "safe operating spaces" for the processes that regulate life support systems on earth.[14] Leaving these safe operating spaces risks triggering ecological tipping points, causing certain ecosystems to shift into an unfavorable state that may be difficult or impossible to reverse.[15] This same concept is used in the field of social transformation as well. If positive social tipping points are triggered, they set off many positive feedback mechanisms, thereby accelerating a transition toward a favorable state.[15]

The theory of change for Social Tipping Points

The degree of system change over time caused by debated social tipping points often takes the shape of an S-curve.[15] This can be seen in Figure 9.2, which illustrates the four-phase pattern of a transformation process.[2,16] The four-phase pattern distinguishes a pre-development phase of slow changes; a take-off phase with more coordinated *niche* activities and *regime* reactions, leading to a social tipping point; an acceleration phase in which structural changes occur; and, lastly, a stabilization phase in which the system stabilizes within the altered state.[2]

In this model, the take-off and acceleration phases are particularly relevant, as they determine whether a tipping point truly activates positive feedback loops. A key question in transition research is how large the population adopting a behavior (social norm) needs to be for it to accelerate.

A behavioral economics study from Damon Centola and colleagues from the universities of Pennsylvania and London that investigated how large a committed minority (the critical mass) needs to be to cross social tipping points has gained popularity in recent years.[17] In their study, Centola and colleagues presented 194 participants with a picture of a person's face and then had them decide on a name for that person. If the participants followed the naming convention of the group (gave the same name as most fellow participants), they were financially rewarded. They were financially punished if they diverged from the naming convention (gave a different name than most fellow participants). Once

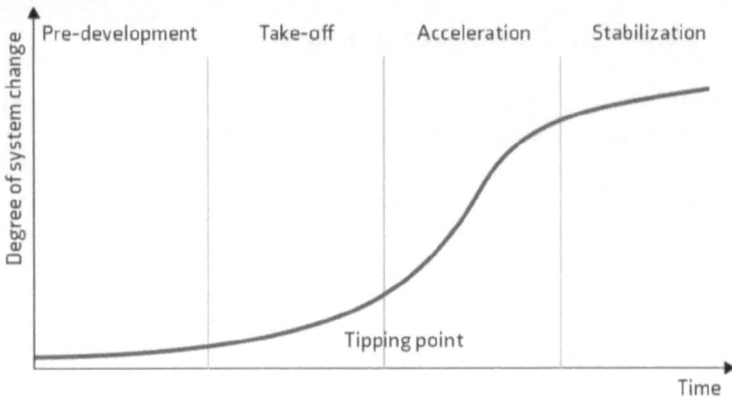

Figure 9.2: S-curve of social tipping points adapted from Göpel[2]

the group was somewhat set on a specific name for that person, the researchers introduced a minority of "new participants" (concealed research assistants), whose secret goal was to overturn the naming convention. For example, they pushed to give a stereotypically female name to a rather male face.

Now, the question remained: How many people would it take for the participants to let go of their established naming convention? The findings of Centola and colleagues showed that the committed minority needed to be around 25% of the size of the group to trigger a social tipping point that overturned the naming convention. Then, 72% up to 100% of all participants switched to the new naming convention in subsequent rounds of the game. This experiment shows that one tipping point of social norms leading to this S-curve can be 25%. It had the advantage that, similar to reality, the game socially and financially encouraged sticking to the old convention at first.[17]

With respect to climate-related conventions, the question is whether 25% of the population adopting individual habits, such as vegetarianism or car-free mobility, could have a critical accelerating effect on transforming societal habits. The scope of this study is inherently limited by its artificial nature. To name one limitation, the experiment deals with freshly formed naming conventions, as opposed to, for example, behavioral mobility conventions that have been established over decades and are stabilized by other routines and infrastructures.

The size of the minority that triggers a tipping point in real life is likely to range in size depending on the *regime* and *landscape* conditions. Jessica Aschemann-Witzel and Maureen Schulze from Aarhus University and Copenhagen Business School emphasize that research on social tipping points is still at an early stage and not yet ready to inform real-world action.[15]

Within the climate movement, and particularly within the climate action group *Extinction Rebellion*, we encounter the idea that "it only takes 3.5% of the population to overthrow the system". This figure is based on an extensive analysis of violent and non-violent social movement campaigns occurring between 1900 and 2006.[18] This analysis, performed by political scientists Erica Chenoweth and Maria Stephan, indicates that when about 3.5% of the whole population is mobilized, social movement success seems inevitable.[19] However, the study itself and its application to the climate context by members of the climate movement have been criticized. For instance, it is questionable whether social transformations away from autocratic or military regimes can be compared to socio-ecological transformations within capitalist democracies.[20]

Regardless of the validity of either of the described claims, it is necessary to build informed theories of change in order to mobilize and not dishearten movement members when change does not happen as rapidly as hoped. According to transition studies, there is not a singular socio-ecological transformation involving a singular *regime*. Rather, we need many socio-ecological transformations in multiple *regimes*, which also require shifts in many social norms. Therefore, it may be less useful to look for and wait for a social movement that is the ideal size and more useful to pay attention to favorable conditions and the interplay of diverse levels, processes, and actors that lead to change.

 Box 9.4: The bottom line

The Social Tipping Point Concept usually promotes a theory of change that takes an S-curve shape. That S-curve goes through four major phases: pre-development, take-off, acceleration, and new stabilization. For newly formed social norms, a minority of 25% could trigger one such tipping point. However, transition researchers suggest that a proper understanding of the conditions of a social tipping point is needed in order to talk about numbers.

THE THREE STRATEGIES OF TRANSFORMATION

In his book *Envisioning Real Utopias*, the analytical Marxist-sociologist Erik Olin Wright introduced a theory of change consisting of three strategies,[3] which can be summarized as *transformation through rupture*, *transformation through creating alternatives*, and *transformation through reform*. But before reflecting on the strategies for transformation, Wright's primary idea was that a theory of social transformation needs to understand the circumstances of transformation (obstacles to change, opportunities for transformation, and long-term trends).

Circumstances of transformation

In contrast to the Multi-Level Perspective, Wright puts an emphasis on obstacles to change: stabilizing forces and power relations within current regimes[3].

Regimes stabilize themselves passively through people's everyday routines and activities. There are also ways in which institutions, such as governments, schools, churches, the courts, and the police actively stabilize and sustain themselves. Methods of such active stabilization can include:

- punishment and repression; for example, raising the costs and risks of collective climate action,
- rules and court proceedings; for example, favoring large, system-conforming corporations over individual claimants in environmental cases,
- the dissemination of ideas; for example, shaping our beliefs about what we want and what socio-ecological ideas we deem possible (in psychology terms: values, norms, and efficacy beliefs), and
- path dependencies; for example, mechanisms by which societal wellbeing depends on successful economic activity.[3]

Though seemingly stable, current systems are also full of gaps and contradictions that provide opportunities for transformation.[3] As society is highly complex and unpredictable, institutions often have inadequate knowledge of how to maintain and change it, make mistakes, and produce unintended side effects that may promote social transformation. Moreover, institutions have not always been designed to fit capitalism, and capitalist societies always face destabilizing moments as they move in circles from regulation to deregulation to re-regulation. These gaps and contradictions create windows of opportunity for social change.

According to Wright, two things are needed for social change to occur: "conditions must be ripe" and there must be a strong strategic movement action.[3] Social movements can help ripen the conditions, but only to a limited extent. Rather, windows of opportunity arise from long-term unintended trends. Social movements must therefore adopt long-term horizons; they must wait for conditions to ripen and detect windows of opportunity in order to be ready to concentrate collective action at exactly those times. This may be somewhat different in the case of the climate crisis, which is likely to produce many ecological disasters that offer more immediate windows of opportunity for collective climate action. Nevertheless,

> the climate movement needs a long-term perspective and strategy that analyzes current beneficial trends while at the same time being prepared for sudden openings in the system.

It is crucial to keep in mind that at the beginning of a crisis, society is often in a state of shock. Some powerful regimes (such as companies, media, political parties) may therefore use windows of opportunity to push through changes that work against socio-ecological change. As well, large protests can have unpredictable consequences and even lead to more repressive policies that further stabilize the status quo (an example of this can be seen in how the regime in Egypt acted following the Arab Spring protests).[21] Wright's interpretation of the process of social transformation is relevant here, as he states that it is neither linear nor S-curved but rather consisting of backlashes and circular processes.[3]

The theory of change for the Three Strategies of Transformation

Wright describes three strategies of social change within economies that range from capitalist to democratic socialist.[3] In this book, we call them *transformation through rupture, transformation through creating alternatives*, and *transformation through reform*. As Wright also notes that not a single one of these strategies is simple and unproblematic, the challenge for social movements is to combine these strategies. Though all three strategies are present in the climate movement, very often they seem to be separated by their ideas and methods.

Transformations through rupture

Transformations through rupture aim to create entirely new institutions and replace whole regimes.[3] Wright calls this idea "smash first, build second".[3] While profound revolutions seem unlikely in contemporary societies, according to Wright, rupture strategies are worth discussing because they can motivate movement members and help us better understand other strategies, and because small-scale ruptures in subsystems might be possible. In a Marxist tradition, Wright compares predicted trajectories of people's material wellbeing, shown in Figure 9.3.

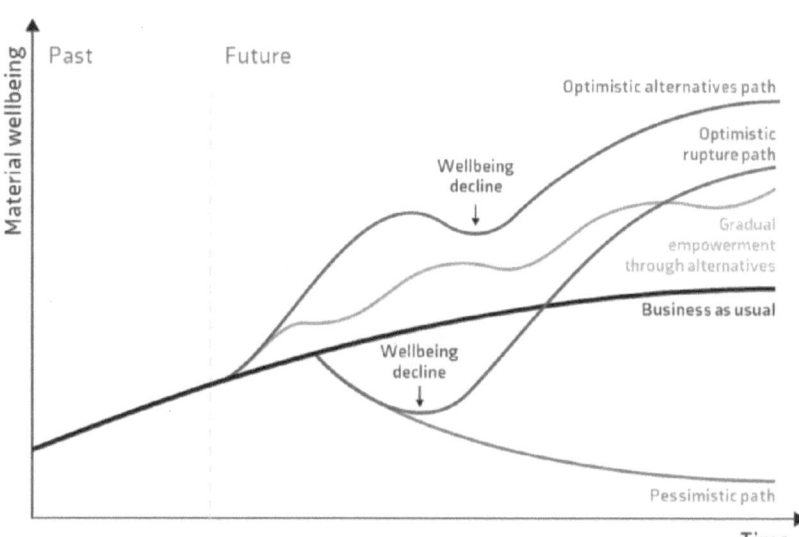

Figure 9.3: Predicted trajectories of material wellbeing in transformation paths, adapted from Wright[3]

Wright assumes that under business as usual conditions the material wellbeing of people in contemporary Western societies will remain constant or slowly increase. When social movements gain large public support for a rupture strategy, rupture becomes more likely. The ideal path for people trying to bring about change assumes that people's wellbeing increases immediately after a rupture. However, Wright suggests that this is not only unlikely but that initially it is actually far more likely to lead to a decrease in wellbeing. In an optimistic rupture path, this decline is temporary, and wellbeing later increases beyond business as usual. However, as wellbeing declines, people are likely to become pessimistic about whether positive change will actually occur in the long run (pessimistic path). In democratic societies, this is likely to lead to a loss of political support in election cycles, even though people may still support the values behind the rupture. This makes *transformations through ruptures* a very implausible strategy within democratic systems.[3]

Wright proposes conditions under which the optimistic pathway is more likely: if the decline in wellbeing is not too deep and prolonged, if people's values are strong enough to sustain them through periods of reduced wellbeing, and if wellbeing is already declining under business as usual,[3] which may indeed wind up being the case as the climate crisis progresses. While large ruptures in democratic societies may be unlikely, the climate movement could try to create favorable conditions for smaller ruptures. For example, the climate movement could focus on building strong values and efficacy beliefs around socio-ecological transformation in the here and now and highlight negative trends in material wellbeing to make drastic socio-ecological changes more appealing to society members.

Transformations through creating alternatives

Transformations through creating alternatives aim to build new forms of socio-ecological living and empowerment in the cracks of the dominant regime, where these alternatives are not threatened by the regime.[3] In the climate movement, we encounter this strategy in community-supported agriculture, eco-villages, and housing cooperatives – these are the places where new socio-ecological structures, ways of living, and togetherness are being tested. Alternatives can be innovations in technology or social sustainability, or both, that alter our relationships with others, ourselves, and nature.[22]

However, the approach of creating alternatives is often criticized for being an escape from reality, for diverting energy and time away from other transformation strategies, and for having to hide within the small spaces the current regimes "allow", thereby not posing a real threat to current power relations.[3] In response to these critiques, Wright highlights how creating alternatives can be useful when combined with other strategies.

First, alternatives can create beneficial conditions for ruptures, leading to an optimistic alternatives path.[3] They can contribute to people's wellbeing, thereby making a brief decline in wellbeing more acceptable. For example, local renewable energy projects can strengthen bonds between community members, making the overall community more resilient in times of crisis.

Pre-established alternatives also make it more likely that innovations will be tested and refined over many years and decades, rather than having to be developed rapidly once a rupture occurs. For example, agro-forestry practices that account for the changing climate may be ready to be implemented on a large scale once a rupture occurs. The pre-testing of alternatives can shorten the transition period from one regime to another and help to stabilize democratic principles in the long run.[3]

Second, alternatives may lead to a softening of some constraints along a path of gradual empowerment.[3] They may modify current power relations step by step, leading to social empowerment and a more profound transformation in the long run. This could be, for example, solidarity housing projects such as the *Mietshäuser Syndikat* [Apartment Complex Syndicate], which withdraws property from the privatized housing market.[23]

Transformations through reform

Transformations through reform aim to symbiotically alter and extend institutions.[3] This strategy is often seen as being at the core of social change in parliamentary democracies and is based on the assumption that transformation will be most stable if it also solves real-world problems within the current regimes. As such, the goal with this strategy is to find positive-sum changes that promote bottom-up social empowerment while simultaneously solving problems of dominant regime actors. But how can we identify these win-win situations?

Drawing on sociological theory, Marie Heitfeld and Alexander Reif from *Germanwatch*, an independent development, environmental, and human rights organization, differentiate between various societal subsystems and their respective system logics and driving interests.[24-26] The logic to which the various subsystems are responsive is also called a "code". The code of the political subsystem is power within a legislative term. The code of the judicial subsystem is law and order across varying terms. The code of the economic subsystem is profit primarily within the short term; specifically, the code of the financial market is profit, opportunity, and risk during a term lasting from a quarter to two years. And the code of the technological subsystem is know-how across varying terms.[24]

An understanding of these codes can help in developing strategies for how actors from the various subsystems can be approached by climate action groups, taking into account their respective interests. The codes also show where opportunities for transformation processes exist (and where they do not) and that the political, financial, and economic subsystems are usually about the short-term.

Many progressive actors share the concern that reform strategies that align with the codes of current regimes will not bring about profound change.[3] However, to prevent the worst (climate) injustices, reforms can be part of larger transformations if they increase the scope for socio-ecological alternatives. For example, the German *Fridays for Future* movement campaigned for reforms to strengthen climate legislation. While the government was then forced by a constitutional complaint to pass a more ambitious climate protection law, it still fell far short of the steps needed. Nevertheless, it was a major decision in that it can be used to provide legal support for future decisions on climate protection. It has also already led to political steps, such as preventing economic subsidies for fossil fuels during the COVID-19 pandemic, more ambitious measures for climate-neutral mobility, and debates on the necessary shift towards renewable heating.

 Box 9.5: The bottom line

Societal systems have mechanisms to stabilize themselves. This is why social movements need to identify the gaps in prevailing regimes, as well as windows of opportunity. Three strategies can support each other on various paths of socio-ecological transformation: *transformation through rupture, transformation through creating alternatives,* and *transformation through reform.*

THE MOVEMENT ACTION PLAN

In his book *Doing Democracy,* social change activist Bill Moyer describes four roles within social change that feed into the broader Movement Action Plan.[4] These roles were also previously mentioned in Focus 2 – Strategy 4 of Chapter 5 on distributing roles. This section features summaries of his hands-on work, as well as historical examples available from the Commons Library.[27-29] Moyer's analysis is based both on experiences from past social movements and on the critique that scientific approaches often focus on social tipping points while neglecting the circular nature of social movements. Moyer's stance is succinctly illustrated in the following quote:

> "While there is much useful information in social movement theories, most do not help us under the ebb and flow of living, breathing social movements as they grow and change over time."[29]

Having noticed that members of social movements were sometimes very frustrated and discouraged at times when they were actually quite successful, Moyer developed the Movement Action Plan in an attempt to explain this phenomenon and to paint the bigger picture.

The four roles within social change

Moyer posits that social movements require four types of organizations and people: *rebels, reformers, change agents* and *citizens.* Let's explore how each of the four roles has different functions within the Movement Action Plan.[28]

Rebels

Rebels typically protest for change and express their critiques of social conditions and policies that violate their beliefs.[27] The climate movement has a long history of *rebels*. Through demonstrations, direct action, and civil disobedience, rebellious activists have been able to temporarily stop climate destruction like coal mining and deforestation.

One example of a *rebel* group acting within the German climate movement is the movement alliance *Ende Gelände*. The strength of their civil disobedience actions is probably that they are able to create lose-lose dilemmas for their targeted opponent: the fossil fuel industry. *Ende Gelände*'s actions leave fossil fuel companies with strictly unfavorable options: they can choose to allow activists to continue physically blocking critical infrastructure, thereby suffering economic damage, or they can use physical force against activists, thereby revealing themselves as active defenders of the grave injustices being exposed by the activists.

According to Moyer, *rebels* are effective when they focus on injustices caused by powerful institutions, attract attention, maintain ties and a shared identity with other groups from the movement, and engage in non-violent protest.[27] *Rebels* become ineffective when they cling too strongly to their moral convictions and resort to violent means out of desperation and a low sense of efficacy.

Reformers

Reformers work on policy changes within political and legal structures and institutions.[27] In the climate movement, they may be government workers, members of green and left parties, and organizations that lobby and advise on successful climate policy.

Reformers are most effective when their aims are in line with the movement as a whole, and when they do not settle for minor steps.[27] It also seems relevant that they maintain ties and shared identities with other movement organizations and roles. This role seems to be the least studied by psychologists.

Change agents

Change agents work towards positive and constructive solutions through engaging, educating, and organizing the public.[27] In the climate movement, we encounter *change agents* from a range of groups, from eco-villages to sustainable university initiatives to organized camps for climate action. They might host a public lecture on sustainable farming or maintain our wind turbines, as shown in Image 9.1.

Change agents are most effective when they aim neither too big (utopia) nor too small (minor steps), promote participatory democracy, connect people inside with those outside the movement, and attend to the needs of movement members; for example, by working to prevent activist burnout.[27]

Image 9.1: Change agent Damien Cuello, who is responsible for the maintenance of the 6-wind turbine facility on Ascension Island.

Photo by Lance Cheung, U.S. Air Force, 2009 (CC BY 2.0)

Citizens

Citizens comprise individuals both inside and outside the movement who support the movement and its cause.[27] This could include *rebels*, *reformers*, and *change agents*, as well as their friends, their family, and many others who support the climate movement and its goals.

Moyer suggests that social movements need large support from *citizens* who are willing to contribute to the success of the movement if they share its values, have political efficacy, and identify with the movement.

The theory of change for the Movement Action Plan

As can be seen in Figure 9.4, the Movement Action Plan includes various phases that our Author Team has grouped together in order to facilitate understanding.

Phases 1–3: Movement creation

In phases 1–3, social movements are formed in politically quiet but unjust times.[28] Initial attempts to change these injustices, such as *reformers* trying to change institutions, fail. Conditions for the movement then begin to ripen. Through growing dissatisfaction with the injustices among *citizens*, constant opposition led by *rebels*, and increased movement organization, people begin to see that change might actually occur. An example of this is the civil rights movement, where the following conditions made the country ripe for social change:

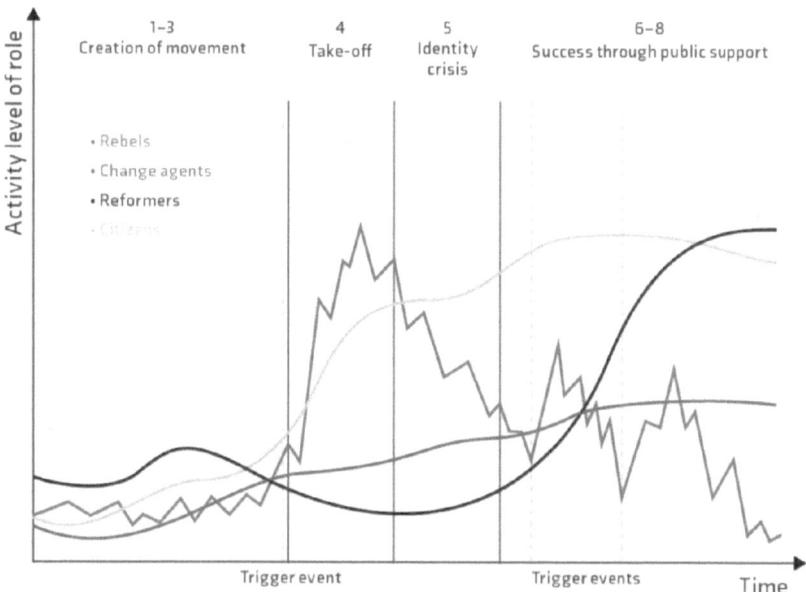

Figure 9.4: Activity level of the four roles in the Movement Action Plan, adapted from Moyer[28]

the emergence of independent Black African countries, the large Northern migration of Blacks who maintained their ties to the segregated South, the rising Black college student population, and the 1954 Supreme Court's Brown vs. U.S. decision [...] provided a legal basis for full civil rights.[28]

Phase 4: Take-off

During phase 4, a trigger event leads to the movement's take-off.[28] Support for the movement and its goals rises sharply among *citizens*, and there is a strong rise in the number of *rebels*. Seemingly overnight, a social injustice becomes the issue everyone is talking about. This phase might offer an explanation for why, historically, social change often seems unexpected and disruptive, even though it is likely that, in most cases, unseen work by *change agents* and *reformers* has already been ongoing. Phase 4 shows that at certain points change can take on a very rapid, non-linear dynamic that is promising but often uncontrollable.

An example of a trigger event is the 1955 arrest of a Black woman named Rosa Parks, who refused to give up her seat in the Whites-only section of a public bus.[28] And an example of "overnight" change is the fall of the Berlin Wall and the German Democratic Republic, which, in reality, came after years of political and economic pressure and near-invisible and suppressed but continuous and growing local protests.

Phase 5: Identity crisis

Interestingly, Moyer then predicts an identity crisis for the movement in phase 5.[28] This may be because of rapid advancement at phase 4 that cannot be maintained in the long term. In phase 5, members of the movement often experience exhaustion from overwork and activist burnout, and some members drop out. At this stage, members feel they have not achieved anything, have had no real victories, and that those in power are simply too powerful.[30] This is accompanied by negative mainstream media coverage and powerful regime agents claiming that the movement is dead.

Here, Moyer sees the possibility of a self-fulfilling prophecy if movement members do not have a theory of change that is capable of explaining this phase. Movement failure is indeed a possibility. However, by recognizing that these feelings are a normal part of the life cycle of social movements, it may be possible to overcome this phase. At this stage, the movement needs to become more strategic, form support groups, and continue to use non-violent strategies.[30] In addition, people and organizations need to understand socio-ecological change as a task that requires not just a temporary action by *rebels* but enduring action by *change agents*.

Phases 6–8: Success through public support

Phases 6–8 are the phases of success through large public support. In these phases, social movements need to see the bigger picture and develop a strategy that will secure public support. This includes continuously raising awareness of the topic and preparing for increased conflict with regime actors.[30] Moyer makes the case that non-violent collective action is needed to win over the support of the public majority.[28] He argues that "social movements involve a long-term struggle between the movement and the power holders for the hearts, minds, and support of the majority of the population"[29] and warns against installing a violent radical flank that could hamper broader support. However, this rejection of a violent radical flank continues to be debated (see Focus 2 – Strategy 4 in Chapter 3 for more on considering the impacts of a radical flank).[31]

If movement activity is sustained, this could lead to success. A trigger event may cause a dramatic showdown, resembling the take-off stage as it mobilizes a necessary amount of citizen support. It could also lead to a quiet showdown in which the prevailing regimes make major concessions. Another possibility could be the slow transformation of the current system over decades through new policies.

Notably, according to the Movement Action Plan, the "success achieved […] is not the end of the struggle but a basis for continuing that struggle and creating new beginnings".[28] The American documentary film *Crip Camp: A Disability Revolution*, for example, illustrates how members of the radical disability rights movement first had to fight for their right to inclusivity and then, in a second wave of protests, press for the implementation of that right. Such accounts show that societal change often requires not just changing one regime but possibly one regime after another, while shifting social norms at the same time (for example, in interactions with people with disabilities).

Phases 6–8 require a lot of activity from *reformers*, the constant engagement of *change agents*, and large support from *citizens*, but relatively less action from *rebels*.

Take-aways

Overall, the Movement Action Plan shows that all roles and inter-role cooperation are essential for a movement's success. It also adds to other theories of change by highlighting that what seems like an unexpected change to some may have been prepared over a long period by others and that possibly not all roles are equally important throughout all phases of socio-ecological transformation. We, the Author Team, do not want to impose our interpretation of which phase the climate movement is currently in. Rather, we suggest using the Movement Action Plan as a tool for reflecting on both individual and climate group actions and goals.

Building on the considerations of the climate action group *ausgeCO2hlt*, different roles within the climate movement need to be better integrated and dynamically adapted to changing conditions, especially when their actions interfere with each other.[22] Indeed, members of the Author Team sometimes experience a disconnect between *rebels* and *change agents*. Some *rebels* may move from one action to the next without building lasting, locally rooted projects. At the same time, some *change agents* may feel they are losing track of the overall struggle as their projects take up their full capacity. Both sides can wind up feeling disconnected and alone.

This is where joint experiences come in, which are important for showing how people in different roles can benefit and learn from each other. For example, a climate action group could occupy fertile farmland not only to protest its destruction (*rebels*), but also to build solidarity agriculture for growing vegetables as an alternative to the current system (*change agents*). Such places could then be secured not only through occupations but also through safeguards within institutions (with the help of *reformers*). This cooperation could generate political pressure through *citizen* support. Taking the hiking metaphor from before, we need to adapt our strategies to the terrain that we're walking through and combine our strengths to reach the destination.

 Box 9.6: The bottom line

The Movement Action Plan distinguishes four roles within social movements: *rebels*, *reformers*, *change agents* and *citizens*. The relevance of each role changes as movements progress through the stages of creation, take-off, identity crisis, and success through public support.

COMMON FEATURES OF THEORIES OF CHANGE

There are clear differences between the theories of change described in this chapter, such as their interpretation of power relations and their focus on technological innovation in capitalist societies or social change through social movements. However, there are some similarities that constitute valuable insights for the climate movement and the design of collective climate action.

Almost all theories consider diverse levels, roles, paths, and strategies, the interplay of which contributes to transformation. Despite originating in different scientific and practical backgrounds, overlaps can be found within such theories of socio-ecological change:[12]

- The circumstances of *landscape* pressure and *regime* destabilization within the Multi-Level Perspective may have some similarities with Wright's *transformations through rupture* strategy or Moyer's *rebel* role.
- When the Multi-Level Perspective highlights the importance of *niche* alternatives, this partially reflects Wright's focus on *transformations through creating alternatives* and Moyer's *change agent* role.
- Geels and Schot's pathway in which a *regime* responds to moderate *landscape* changes and adapts by including *niche* innovations shares some overlap with Wright's *transformations through reform* strategy and Moyer's *reformer* role.

Other common features of some of these theories are that they see socio-ecological change as non-linear, sometimes seemingly unexpected but actually prepared, and dependent on external forces that cannot be influenced. This makes transformations only controllable to a limited extent. Climate action groups should acknowledge and reflect on these features in order to create effective collective climate action that feeds into greater transformation.

YOUR TAKE ON SOCIO-ECOLOGICAL CHANGE

This chapter was indeed a leap into the bigger picture. Now, you get the chance to paint your own picture. With the help of the following questions, you can individually brainstorm or collectively discuss your own engagement in collective climate action with respect to the introduced levels, tipping points, strategies, and roles that are important in socio-ecological transformations (adapted from Heitfeld and colleagues[32]):

- *If you are involved in a climate action group, which role do you currently see yourself occupying? What are your strengths in this role? What are the limits of this role?*
- *Which organizations, groups, or other actors do you think fill other important roles in your field of action?*
- *Which levels, strategies, and roles do you think are most effective? Why?*
- *What could be crucial tipping points?*
- *What stage of transformation do you think the climate movement is in? What about the movement in your specific area of action?*

- *Which roles have been under-occupied so far but would be very relevant right now?*
- *Do you feel the need to adjust your level, strategy, or role in order to work even more effectively towards your goals? If so, what do you need to do to achieve this?*

References

1. Geels, F. W. & Schot, J. Typology of sociotechnical transition pathways. *Res. Policy* 36, 399–417 (2007). https://doi.org/10.1016/j.respol.2007.01.003
2. Göpel, M. *The Great Mindshift*. vol. 2 (Springer International Publishing, 2016). https://doi.org/10.1007/978-3-319-43766-8
3. Wright, E. O. *Envisioning Real Utopias*. (Verso, 2010).
4. Moyer, B. *Doing Democracy: The MAP Model for Organizing Social Movements*. (New Society Publishers, 2001).
5. Seyfang, G. & Haxeltine, A. Growing grassroots innovations: Exploring the role of community-based initiatives in governing sustainable energy transitions. *Environ. Plan. C Gov. Policy* 30, 381–400 (2012). https://doi.org/10.1068/c10222
6. WBGU Welt im Wandel: Gesellschaftsvertrag für eine Große Transformation. *Zusammenfassung für Entscheidungsträger*. (Wiss. Beirat der Bundesregierung Globale Umweltveränderungen, 2011).
7. Wullenkord, M. C. & Hamann, K. R. S. We need to change: Integrating psychological perspectives into the multilevel perspective on socio-ecological transformations. *Front. Psychol.* 12, 655352 (2021). https://doi.org/10.3389/fpsyg.2021.655352
8. I.L.A. Kollektiv. *Die Welt auf den Kopf stellen*. (oekom Verlag, 2022).
9. Göpel, M. & Remig, M. Mastermind of system change: Karl Polanyi and the "Great Transformation" [Vordenker einer nachhaltigen Gesellschaft. Karl Polanyi und die „Große Transformation"]. *GAIA - Ecol. Perspect. Sci. Soc.* 23, 70–72 (2014). https://doi.org/10.14512/gaia.23.1.19
10. Manstetten, R., Kuhlmann, A., Faber, M. & Frick, M. Grundlagen sozial-ökologischer Transformationen: Gesellschaftsvertrag, Global Governance und die Bedeutung der Zeit. ZEW – Centre for European Economic Research Discussion Paper No. 21-034 (2021). https://doi.org/10.2139/ssrn.3824656
11. Jessop, B. *Putting Civil Society in Its Place: Governance, Metagovernance and Subjectivity*. (Policy Press, 2020). https://doi.org/10.56687/9781447354970
12. Geels, F. W. The multi-level perspective on sustainability transitions: Responses to seven criticisms. *Environ. Innov. Soc. Transit.* 1, 24–40 (2011). https://doi.org/10.1016/j.eist.2011.02.002
13. Köhler, J. *et al.* An agenda for sustainability transitions research: State of the art and future directions. *Environ. Innov. Soc. Transit.* 31, 1–32 (2019). https://doi.org/10.1016/j.eist.2019.01.004
14. Rockström, J. *et al.* A safe operating space for humanity. *Nature* 461, 472–475 (2009). https://doi.org/10.1038/461472a
15. Aschemann-Witzel, J. & Schulze, M. Transitions to plant-based diets: The role of societal tipping points. *Curr. Opin. Food Sci.* 51, 101015 (2020). https://doi.org/10.1016/j.cofs.2023.101015
16. Mersmann, F., Wehnert, T., Göpel, M., Arens, S. & Ujj, O. *Shifting Paradigm: Unpacking Transformation for Climate Action; A Guidebook for Climate Finance & Development Practitioners*. https://epub.wupperinst.org/frontdoor/index/index/docId/5518 (2014).

17. Centola, D., Becker, J., Brackbill, D. & Baronchelli, A. Experimental evidence for tipping points in social convention. *Science* 360, 1116–1119 (2018). https://doi.org/10.1126/science.aas8827

18. Chenoweth, E. & Stephan, M. J. *Why Civil Resistance Works: The Strategic Logic of Nonviolent Conflict.* (Columbia University Press, 2011).

19. Robson, D. The '3.5% rule': How a small minority can change the world. *BBC.* https://www.bbc.com/future/article/20190513-it-only-takes-35-of-people-to-change-the-world (2019).

20. Matthews, K. R. Social movements and the (mis)use of research: Extinction Rebellion and the 3.5% rule. *Interface* 12, 591–615 (2020).

21. Engler, M. & Engler, P. *This is an Uprising: How Nonviolent Revolt is Shaping the Twenty-First Century.* (PublicAffairs, 2016).

22. ausgeCO2hlt. *Jenseits von Hoffnung & Zweifel: Gedanken zum Widerstand in der Klimakrise.* (UNRAST, 2022).

23. Mietshäuser Syndikat. Mietshäuser Syndikat – Die Häuser denen, die drin wohnen. https://www.syndikat.org/

24. Heitfeld, M. & Reif, A. Transformation gestalten lernen. https://www.germanwatch.org/de/19607 (2021).

25. Luhmann, N. *Die Gesellschaft der Gesellschaft. 2.* (Suhrkamp, 2021).

26. Habermas, J. *Faktizität und Geltung: Beiträge zur Diskurstheorie des Rechts und des demokratischen Rechtsstaats.* (Suhrkamp, 2019).

27. Moyer, B. The four roles of social activism. *The Commons.* https://commonslibrary.org/the-four-roles-of-social-activism/ (2001).

28. Rose, A. Bill Moyer's Movement Action Plan. *The Commons.* https://commonslibrary.org/resource-bill-moyers-movement-action-plan/ (2012).

29. Engler, M. & Engler, P. Surviving the ups and downs of social movements. *The Commons.* https://commonslibrary.org/surviving-the-ups-and-downs-of-social-movements/ (2014).

30. Moyer, B. Bill Moyer – The Movement Action Plan. https://www.historyisaweapon.com/defcon1/moyermap.html (1987).

31. Malm, A. *How to Blow up a Pipeline: Learning to Fight in a World on Fire.* (Verso, 2021).

32. Heitfeld, M., Baum, D. & Hildebrand, L. *Dein Handabdruck für die Agrarund Ernährungswende: Ein Do-It Guide zum Loslegen.* (2021).

10

EFFECTIVE **GOALS**
FOR **CLIMATE** GROUPS

DOI: 10.4324/9781003558439-12

PICKING YOUR BATTLES: GOALS AND TARGET GROUPS

As daunting a task as it is to confront the systems that are exploiting humans and nature, time and again moments of progress emerge in these struggles.

In the United States in 2021, a group of people prevented the construction of a pipeline that would have affected a neighborhood of predominantly Black people in the southern state of Tennessee. One of the group's organizers described their success by explaining that

> when the voices of those most excluded and marginalized become the leading voices in advocating environmental justice, it ultimately benefits everyone.[1]

The organizer also mentioned that, in addition to stopping the construction of a pipeline in Memphis, they truly ignited a broader movement for social justice and change.

That same year, students at Harvard pushed the university into divesting from fossil fuels, and environmental lawyers won a case that legally bound Shell to drastically reduce its emissions.[1] Have you ever experienced a moment of victory as part of the climate movement or a climate action group? Some successes are big, some successes are small. Some successes are instant, some only realized years after.

If climate action groups are to be effective in contributing to socio-ecological change, they need to discuss and define success. And they need to consider how to measure the impact of collective climate action. Often, the ultimate aim is a socio-ecological transformation that leads to something similar to what was formulated in the Brundtland Report by the World Commission on Environment and Development of the United Nations: a good life for all forever within the planetary boundaries.[2]

However, as you have seen, there are many different strategies and small-to medium-sized goals that can be pursued to achieve the overall aim of a socio-ecological transformation. Which roles do you want to take on, which strategies do you want to adopt, and which theory of change do you want to follow? Is your goal to increase the number of people willing to join protests? Or is it to convince your mayor to implement the demands of your neighborhood initiative? Is your goal to keep the road blocked for a certain amount of hours? Is it attracting public attention or gaining support within your local community? Or is your goal a major change in legislation?

When we as the Author Team ask climate action groups about their goals, the answers range from precise goals to "it depends" to "of course, we want to gain public attention and support, and achieve numerous further goals all at once". It is important for a group to be aware of their strategy, with clearly defined immediate, short-, medium-, and long-term goals, so that it can effectively work towards socio-ecological change.

Luckily, Robyn Gulliver and colleagues from the University of Queensland have proposed a framework for effective activism that can help climate action groups in this endeavour.[3] This framework may build a bridge between larger theories of socio-ecological transformation and planning actions in a specific climate action group. The framework is not empirically tested and does not claim to be exhaustive. Rather, it exists as an adaptable guide to help individuals choose which goals to prioritize in their climate action group and in a specific planned collective climate action. This framework distinguishes five target groups: *Self, Supporters, Bystanders, Third Parties*, and *Opponents*.[3]

Your personal *Self* immediately seeks affirmation, empowerment, meaning, and solidarity and wants to have emotional experiences. In the short- to medium-term, *Self* may want to build and maintain friendships in climate action groups, feel secure, and sustain their participation.

Supporters are members of specific climate action groups as well as those individuals and groups who support the socio-ecological goals of others. This could be a supportive friend or a trade union with which a climate action group is positively connected. Climate action groups can pursue various goals that address *Supporters*. As an immediate goal, they may want to affirm their group identities, let them express group values, and empower them, while short- and medium-term goals may be to generate intentions, initial actions, and sustained action.[3]

Bystanders are neutral individuals, such as neighbors who don't have ties to climate action groups, and *Third Parties* are groups with their own priorities and agendas, such as schools. Immediate goals in addressing these two target groups could include raising awareness, building sympathy, and creating shared identities, while in the short- to medium-term, a climate action group may want to build coalitions, generate intentions and actions, and avoid counter-mobilization that might turn them into opponents.[3]

Opponents are those individuals and groups who are actively working against socio-ecological change; for instance, climate deniers and fossil fuel lobbyists. A climate action group may want to reject their values immediately and affirm that there is opposition to them. Short- to medium-term goals could be avoiding counter-mobilization and converting, provoking, or diverting the opposition group and its members.[3]

Gulliver and colleagues' framework has been redesigned into a table to emphasize the parts most relevant to our readers. This table provides an overview of the different target groups and the goals you and your climate action group might want to pursue for each. Following this table, we have also provided additional information and strategies to help you and your climate groups achieve your goals and objectives. Each strategy is supplemented with a reference to the relevant chapter in this book so that you can easily go back and find more information. Let's take a look at Tables 10.1 to Table 10.7, which summarize key messages of this book.

TARGET GROUP DESCRIPTIONS

Table 10.1: Target group descriptions

Target group	Who is this?	Examples	Short- and medium-term goals	Long-term goals
Self	You	You, including your thoughts, feelings, and actions	• Satisfying your need for belonging, meaning, high self-esteem • Being part of a group that nurtures you • Perceiving efficacy and contributing • Influencing others positively • Sustaining action • Having fun • Feeling supported • Having friends • External needs like security, resources, and status	Systemic socio-ecological transformation
Own group	Your climate action group	Members of your group, your group dynamics	• Promoting feeling part of and being attached to the group • Creating a joint moral foundation • Empowering group members • Fostering collective action • Sustaining action • Choosing roles within the movement • Affecting change	Systemic socio-ecological transformation
Supporters	People and groups who support your socio-ecological goals	Friends, feminist groups	• Building connections to the climate action group • Being part of one movement • Supporting one another • Empowering supporters • Fostering collective action • Creating public support • Choosing roles within the movement • Sustaining action as one larger movement	Systemic socio-ecological transformation

Table 10.1: (Continued)

Target group	Who is this?	Examples	Short- and medium-term goals	Long-term goals
Bystanders/ third parties	Neutral people and groups with their own agendas and no ties to climate action groups	Neighbors, schools	• Raising awareness • Building sympathy • Creating shared identities • Fostering collective action • Building coalitions • Avoiding counter-mobilization	Systemic socio-ecological transformation
Opponents	People and groups actively working against socio-ecological goals	Climate deniers, fossil fuel lobbyists	• Rejecting opponents' values • Strengthening opposition • Avoiding counter-mobilization • Appeasing, conciliating, converting, provoking or diverting opponents	Systemic socio-ecological transformation

Source: Adapted from the original by Gulliver and colleagues[3] and supplemented by an additional group: one's own climate action group.

STRATEGIES FOR ACHIEVING TARGET GROUP GOALS

Achieving these goals using SOCIAL IDENTIFICATION (Chapter 2)

Table 10.2: Achieving goals using social identification

For yourself	• Seek experiences in your group that make you feel competent, like you belong, are esteemed and that give you meaning. Help other members do the same • Cultivate solidarity within your group • Let yourself have fun • Look for situations in which you can perceive many people joining climate action
In your group	• Make it possible to identify with your group prototype and leaders • Have experiences together to form a bond • Make visible the fact that social norms are shifting towards climate action and that many people are already engaged in it • Cultivate solidarity within your group • Reflect on how to include marginalized or discriminated groups • Be welcoming and benevolent • Do fun activities together • Help members find their purpose in life • Build a shared understanding of what your group stands for • Give your group unique features • Highlight that your group is valued within the climate movement • Give your members autonomy in their tasks

Table 10.2: (Continued)

For supporters	• Highlight a common fate and common grievances • Make it possible to identify with your group prototype and leaders • Link the supporters and their social groups to climate action • Highlight action-promoting social norms • Create an overarching group • Engage in constructive dialogue • Have experiences together • Make visible the fact that social norms are shifting towards climate action and that many people are already engaged in it • Be welcoming and benevolent • Formulate critique of other groups in solidarity
For bystanders/ third parties	• Highlight a common fate and common grievances • Make it possible to identify with your group prototype and leaders • Link the bystanders' social groups and third parties to climate action • Highlight action-promoting social norms • Create an overarching group • Engage in constructive dialogue • Go through an experience together • Use value-based communication • Use the block leader approach • Increase the visibility of pro-climate norms and trends • Highlight that your group and social movements are valuable
For opponents	• Consider how peaceful protests can escalate • Link the opponents' social groups to climate action • Use value-based communication • Create an overarching group

Achieving these goals using MORAL BELIEFS (Chapter 3)

Table 10.3: Achieving goals using moral beliefs

For yourself	• Reflect on your own values • Let yourself experience anger • Talk about your outrage • Choose your strategies with the radical flank effect and impacts of constructive disruption in mind
In your group	• Reflect on your group's values • Draw attention to injustices • Show who is responsible • Let your members talk about their outrage • Choose your strategies with the radical flank effect and impacts of constructive disruption in mind

Table 10.3: (Continued)

For supporters	• Draw attention to injustices • Show who is responsible • Communicate your outrage to other supporters • Challenge but don't threaten others' moral self-image • Consider the impacts of a radical flank • Make it so your climate action is perceived as legitimate, relatable, and effective by supporters
For bystanders/ third parties	• Consider the activist dilemma • Draw attention to injustices • Show who is responsible • Communicate your outrage • Seek direct contact in everyday life to narrow the moral-empathy gap • Seek direct contact during an action • Challenge but don't threaten others' moral self-image • Use *non-violent communication* • Balance public sympathy and media attention • Consider the impacts of a radical flank • Make it so your climate action is perceived as legitimate, relatable, and effective
For opponents	• Seek direct contact in everyday life to narrow the moral-empathy gap • Consider the impacts of a radical flank • Make it so your climate action is perceived as legitimate, relatable, and effective

Achieving these goals using FRAMING (Chapter 4)

Table 10.4: Achieving goals using framing

For yourself	• Choose which stories you want to tell • Reflect on which frames suit you best
In your group	• Choose which stories you want to tell as a group • Reflect on which frames suit your group and target group best • Consider that you'll have members who prefer a promotion and prevention frame
For supporters, bystanders/ third parties/ opponents	• Identify problems and attributions (diagnostic frames) • Predict the likely course of events (prognostic frames) • Promote the desire to achieve something (motivational frames) • Include a promotion and prevention frame • Frame climate issues locally • Frame climate issues in terms of present and future losses • Frame climate issues as an environmental threat but also as an economic, health, and national security threat • Frame climate issues in terms of people's existing values

Achieving these goals using EFFICACY BELIEFS (Chapter 5)

Table 10.5: Achieving goals using efficacy beliefs

For yourself	• Seek environments that highlight success and the efficacy of your action • Consider your media consumption and linked variations in efficacy beliefs • Remind yourself of any sudden successes • Attribute a success also to your own contribution • Seek to give and get appreciative feedback • Create your own visions of a socio-ecological future (for example, with the aid of literature or movies) • Reflect on your optimal group size and discuss it with your climate action group • Reflect on what goals you want to pursue and how achievable they should be to motivate yourself • Set a small number of diverse goals (include a goal safety net) • Reflect on your own skills • Look for support to develop new skills (including those out of your comfort zone) • Reflect on the roles that you want to take in your climate action group and the climate movement
In your group	• Highlight a successful action and its effectiveness • Remind yourself of any sudden successes • Tell each other success stories • Attribute a success to your own group • Build an appreciative social feedback culture • Design collective climate actions that inspire people and make them feel enthusiastic, hopeful, proud, and moved by them • Create visions of a socio-ecological future as a group • Build a group culture around different motivations • Consider your group size • Reflect on what goals you want to pursue and how achievable they should be • Set a small number of diverse goals (include a goal safety net) • Familiarize yourself with the skills that are present in your group • Support each other in developing skills • Reflect and distribute roles within your group • Consider that well-known and comfortable roles may not lead to the most progress
For supporters	• Highlight a successful action and its effectiveness, especially one where the success was immediate • Give other groups appreciative feedback • Combine social norms with practical ways of how one can actually contribute • Highlight an incidental injustice problem as a hook • Design collective climate actions that inspire people and make them feel enthusiastic, hopeful, proud, and moved by them • Create your joint visions as groups of the climate movement • Reflect on your goals with other groups • Consider that your group may have skills that other groups need (and vice versa) • Reflect on roles and distribute them wisely within the movement

Table 10.5: (Continued)

For bystanders / third parties	• Highlight a successful action and its effectiveness, especially one where the success was immediate • Make visible the fact that social norms are shifting towards climate action and that many people are already engaged in it • Combine social norms with practical ways of how one can actually contribute • Focus on normative rather than non-normative forms of protest • Highlight an incidental injustice problem as a hook • Design collective climate actions that inspire people and make them feel enthusiastic, hopeful, proud, and moved by them • Provide visions of what a socio-ecological future could look like
For opponents	• Show that many people are already engaging in collective climate action

Achieving these goals using COLLECTIVE ACTION (Chapter 6)

Table 10.6: Achieving goals using collective action

For yourself	• Participate in collective climate action; this can foster motivation • Acknowledge the intention-behavior gap • Successful actions: celebrate yourself • Unsuccessful actions: seek group activities that help restore your energy; build up your efficacy beliefs; reframe goals and the setback; take a break and distract yourself
In your group	• Perform collective climate actions together • Acknowledge the intention-behavior gap • Pick your battles • Successful actions: celebrate yourself, evaluate your collective climate action • Unsuccessful actions: engage in constructive dialogue about strategy (changes), restore members' sense of belonging, foster members' efficacy beliefs, reframe goals and the setback
For supporters	• Invite friends and family to collective climate actions • Successful actions: celebrate with others, discuss next steps with other groups • Unsuccessful actions: seek advice from supporters, establish solidarity
For bystanders/ third parties	• Invite friends and family to collective climate actions • Acknowledge the intention-behavior gap

Achieving these goals using RESILIENCE STRATEGIES (Chapter 8)

Table 10.7: Achieving goals using resilience strategies

For yourself	• Make time for recreation • Reflect on the right balance of motivation • Blow off activist steam • Increase your feelings of efficacy • Say no, spread tasks, trust others • Maintain non-activist networks • Consider therapy • Switch to a different type of action • Refrain from any group actions for a while • Choose groups that provide you with your required basis of resources and security
In your group	• Ensure members feel a sense of belonging • Take time to celebrate your group successes • Create fun moments • Let emotions shine through • Develop an organized conflict culture • Reflect on group cultures of martyrdom, competition, and performance • Check your resources • Make healthy engagement an active subject of discussion
For supporters	• Reflect on resilient collective climate action with other groups • Be kind to other groups within the movement

DEFINING YOUR GOALS AND STRATEGIES

With these tables in mind, you can now figure out more precise goals and strategies that your climate action group wants to pursue in order to contribute to a larger socio-ecological transformation. For your group to reach its fullest potential, it is worth discussing these questions:

Clarifying your long-term goal

- *What is your group's long-term focus, vision, or mission?*
- *What is your theory of change?*
- *What is your group's role in the movement for a socio-ecological transformation, and what is currently needed?*

Picking your battles

- *What goals do you want to achieve?*
- *Which of these goals are immediate, short-term, and medium-term?*
- *Which categories do these goals fit into? More general categories may be building awareness, building sympathy, getting people to act, sustaining engagement, building coalitions, and avoiding opponents.*[3]
- *Which goals are most important for your group?*

- *Which goals are less important for your group?*
- *Are there any overlooked goals that might be worth considering?*
- *Which target groups do you think are most important to reach?*

Reflecting on your chosen goals

- *What types of collective action can promote your goals?*
- *What psychological strategies can help you achieve your goals?*
- *What other groups are there addressing the same goals in similar or different ways?*
- *Are there any conflicting goals so that pursuing one goal actually prevents the achievement of another goal?*
- *Will your goals have any adverse effects on other climate actions or target groups?*

You can use these questions to gain a personal understanding of where your climate action group is heading and whether this matches your ambitions and theory of change. In a climate action group meeting, start with an open brainstorming session in which group members can share any goals that come to mind. These could be clustered into immediate, short-term, and medium-term goals by giving them the same color or dividing them into columns. No criticism should be allowed at this stage.

When the brainstorming has ended, you can start prioritizing the goals. *Wandelwerk* likes to give each member a certain number of points to allocate to specific goals. For example, they could give their three points to one goal that they think is most important or give one point each to three different goals. This will give you an overview of the goals people prioritize within your group.

Now, reflect on what actions may achieve these goals. Keep in mind that it is impossible to plan collective climate action that will meet all your goals at once. An action may bring you closer to some goals, while others may not, and some cannot be pursued simultaneously. A more radical type of action may fulfill your goal of attracting more public attention but could negatively affect your goal to achieve public support for the climate action group (consider the activist's dilemma in Focus 2 – Strategy 3 of Chapter 3).

As your climate action group does not exist in an isolated vacuum but in an ecosystem of other actors and groups, your group's actions are likely to have an impact on other groups and possibly on the movement as a whole (see Focus 2 – Strategy 4 in Chapter 3 for more on considering the impacts of a radical flank). It seems wise to think carefully about the impact of your actions on the movement and to take care of your allies. But not just your allies. Actions can affect each of the five target groups differently. When planning an action, it is therefore worth considering how it might affect each target group.

In the final step, you can use the psychological strategies and suggestions outlined in Tables 10.2 to 10.7 to plan and refine your collective climate actions. We, the Author Team, hope that this framework, these questions, and these steps will help you and many climate action groups to design collective climate actions effectively and resiliently.

> ☼ **Box 10.1: Food for thought – What did you discover?**
>
> On the first pages of this book (in Box 1.2), you may have come up with a couple of questions about the psychology of collective climate action – now is the time to reflect back on these questions. What answers have you found, and what answers do you still have left to discover?

References

1. Uyeda, R. L. The climate victories of 2021 that put fossil fuels in check. *The Guardian.* https://www.theguardian.com/environment/2021/dec/31/climate-victories-2021-activism-shareholder-rebellions (2021).
2. United Nations. *Our common future: Report of the world commission on environment and development.* https://www.are.admin.ch/are/en/home/media/publications/sustainable-development/brundtland-report.html (1987).
3. Gulliver, R., Wibisono, S., Fielding, K. S. & Louis, W. R. *The Psychology of Effective Activism.* (Cambridge University Press, 2021). https://doi.org/10.1017/9781108975476

11

CONCLUSION, RECOMMENDATIONS, AND FURTHER IDEAS

DOI: 10.4324/9781003558439-13

FINAL WORDS

The Author Team would like to thank everyone who has contributed to this book in one way or another. For reflecting on and discussing the issues we have raised and for considering psychology-based advice, we would also like to thank both those already involved and those just getting started in the movement for a socio-ecological transformation. The collective of psychologists here at *Wandelwerk* does more than write books and pen articles; our primary means of reaching people is actually through keynotes, workshops, and (university) seminars. We also enjoy engaging in counseling and giving interviews.

If you're interested in working with us, please email us at info@wandel-werk.org, check out our website at www.wandel-werk.org/en, or contact our authors directly at the following email addresses:

Karen: karen.hamann@wandel-werk.org
Eva: eva.junge@wandel-werk.org
Paula: paula.blumenschein@wandel-werk.org
Sophia: sophia.dasch@wandel-werk.org
Alex: a_wernke@yahoo.de
Julian: julianbleh@posteo.de

RESEARCH AREAS IN NEED OF ADDRESSING

As part of the EPEAC project of which this book is part, the Author Team also conducted a study interviewing members of the movement for a socio-ecological transformation about the psychological questions they have and challenges they face.[1] Based on these interviews and our summary of previous psychological research found in this book, we would like to highlight a number of research gaps as can be seen in Table 11.1. If you end up addressing these gaps during your own research, we would appreciate hearing from you.

Table 11.1: Research areas in need of addressing

Motivating collective climate action	• How can climate action groups foster need fulfilment among their members? • How can we confront structural barriers that lead to unequal access to collective climate action? • What roles do hatred and contempt play in predicting involvement in (non)normative collective climate action?[a, b] • How do radical flanks affect the movement for a socio-ecological transformation? • How do people perceive climate-related constructive disruption?

Table 11.1: (Continued)

	• Which narratives should climate groups use for their actions? • What are truly effective ways to increase efficacy beliefs? • What is the best way to deal with goals in the face of victory or defeat? • Which roles in the movement build on which motivation? • What motivates reformers to engage in collective climate action? • Which theories of change can motivate collective climate action?
Resilient collective climate action	• What predicts activist burnout? • Does feeling that climate action is one's core purpose in life act as a buffer against activist burnout? • How can climate action groups deal with anger in a motivating and constructive way? • How can climate action groups deal with traumatic climate action experiences and repression? • What group methods could increase the sharing of emotions?
Effective collective climate action	• What is psychology's role in a socio-ecological transformation and transition research? • What are the distinguishable roles within movements and how effective are they? • How can groups be organized effectively, especially if these have flat hierarchies and anarchist structures? • What effects do certain types of identity, moral, and efficacy communication have on opponents and counter-mobilization? • What effects do certain collective climate actions have on opponents and counter-mobilization?

a. Tausch, N. *et al.* Explaining radical group behavior: Developing emotion and efficacy routes to normative and nonnormative collective action. *J. Pers. Soc. Psychol.* **101**, 129–148 (2011). https://doi.org/10.1037/a0022728

b. Shuman, E., Cohen-Chen, S., Hirsch-Hoefler, S. & Halperin, E. Explaining normative versus nonnormative action: The role of implicit theories. *Polit. Psychol.* **37**, 835–852 (2016). 210

RECOMMENDED BOOKS AND VIDEOS TO EXPLORE

Of course, this book is not the only valuable resource on the psychology of collective climate action. If you want to explore this topic further, Table 11.2 lists several books and videos that we recommend.

Table 11.2: Recommended books and videos

For motivating collective climate action	*Social Change Lab: Psychology of Change: Video Resources and Evidence-based Strategies to Create Social Change*
For resilient collective climate action	*The Lifelong Activist: How to Change the World without Losing Your Way*[a] *Active Hope: How to Face the Mess We're in without Going Crazy*[b] *Mutual Aid: Building Solidarity During This Crisis (and the Next)*[c]
For effective collective climate action	*The Psychology of Effective Activism*[d] *Reimagining Activism: A Practical Guide for the Great Transition*[e]
For climate communication	*Don't Even Think About It: Why our Brains are Wired to Ignore Climate Change*[f]
For private-sphere climate action	*Psychology of Environmental Protection – Handbook for Encouraging Sustainable Actions*[g]

a. Rettig, H. *The Lifelong Activist: How to Change the World Without Losing Your Way.* (Lantern Books, 2006).
b. Macy, J. & Johnstone, C. *Active Hope: How to Face the Mess We're in without Going Crazy.* (New World Library, 2012)
c. Spade, D. *Mutual Aid: Building Solidarity During This Crisis (and the Next).* https://theanarchistlibrary.org/library/dean-spade-mutual-aid
d. Gulliver, R., Wibisono, S., Fielding, K. S. & Louis, W. R. *The Psychology of Effective Activism.* (Cambridge University Press, 2021). https://doi.org/10.1017/9781108975476
e. Narberhaus, M. & Sheppard, A. *Reimagining Activism: A Practical Guide for the Great Transition.* (Smart CSOs Lab, 2015).
f. Marshall, G. *Don't Even Think About It: Why Our Brains are Wired to Ignore Climate Change.* (Bloomsbury USA, 2015).
g. Haman n, K., Löschinger, D. & Baumann, A. *Psychology of Environmental Protection – Handbook for Encouraging Sustainable Actions.* (2016) www.wandel-werk.org/en/materialien

Reference

1. Hamann, K. R. S. *et al.* How can psychological research support movements for socio-ecological change? A qualitative study on psychological challenges and questions of activists. *Global Environmental Psychology.* https://psycharchives.org/en/item/2723d856-0b7b-459c-a79f-f51c7ea529e0 (2024)

Appendix
OVERVIEW OF RESEARCH DESIGNS

INDIVIDUAL EXPERIENCES

Our own experiences are oftentimes very useful for illustrating certain topics. While the Author Team is happy to share their experiences with you, please keep in mind they may be different from your own or those of the groups you belong to. Furthermore, our Team's experiences are predominantly characterized by a White and academic perspective of the Global North.

QUALITATIVE RESEARCH

Qualitative research is present in psychology but much more common in other scientific disciplines such as sociology or political science. Here, the focus can be general or specific case studies that highlight a very unique phenomenon, as shown in Table A.1.

QUANTITATIVE RESEARCH

Psychological research is mostly quantitative, based on data from large numbers of people, and it is often somewhat generalizable. As you can see in Table A.2, research can be viewed as a weighing task between causality inference and reality inference. Often, studies mix some of the described designs. If you are interested in a more extensive overview, take a look at pages 87–90 of *Applying Social Psychology*[1].

Table A.1: Example of a qualitative research design

Design	Description	Causality inference	Reality inference	Typical aim
Interview studies	• research based on interviewing people about a particular topic • can take various forms of analysis • example: assessing empowerment during and after an anti-roads occupation[2]	• self-reported causality • perceptions of interviewees • interpretation of the scientist	• close to real life	• depends on context

Table A.2: Quantitative research methods

Design	Description	Causality inference	Reality inference	Typical aim
Experimental research	• participants are divided into different groups and tested to see how these groups differ in the expression of psychological characteristics after being given a text, task, or other variation between conditions • typically conducted in a laboratory or online • example: comparing an experimental group who receives a message highlighting the efficacy of a group with a control group who receives no such message in order to see how collective action intentions are affected[3]	• causality test	• often far from real life	• studying psychological mechanisms
Longitudinal research	• measures psychological processes across two or more time points • example: measuring people's collective actions and private behaviors and how they change across four time points in order to assess how these concepts influence each other[4]	• hints of causality	• depends on context	• studying psychological mechanisms
Field intervention research	• tests an intervention that is applied to a real-life setting • can include a control group or a measure point before and after the intervention • example: testing the effects of a twelve-week video intervention on people's collective climate action[5]	• depends on research design	• close to real life	• generating answers to real-world problems

Cross-sectional research	• measures concepts at one time point and looks at their relation • example: assessing the relation between need fulfilment and activist burnout[6]	• no causality can be inferred	• depends on context	• depends on context • typically a first step for new areas of research
Meta-analytical research	• includes and statistically integrates the (often quite diverging) findings of a number of related studies • example: a meta-analysis investigating the psychological antecedents of collective action across various studies[7]	• depends on research design • yes, if experimental results are summarized • no, if cross-sectional results are summarized	• depends on context • yes, if field research is summarized • no, if laboratory studies are summarized	• gaining an overview • creating more generalizable insights

References

1. Buunk, B. & Van Vugt, M. *Applying Social Psychology: From Problems to Solutions.* (SAGE, 2013).
2. Drury, J. & Reicher, S. Explaining enduring empowerment: A comparative study of collective action and psychological outcomes. *Eur. J. Soc. Psychol.* 35, 35–58 (2005). https://doi.org/10.1002/ejsp.231
3. Jugert, P. *et al.* Collective efficacy increases pro-environmental intentions through increasing self-efficacy. *J. Environ. Psychol.* 48, 12–23 (2016). https://doi.org/10.1016/j.jenvp.2016.08.003
4. Hamann, K. R. S. *Psychological Empowerment in the Context of Environmental Protection: How Can Personal, Collective, and Participative Efficacy Beliefs Foster Proenvironmental Behavior and Activism?* (University Koblenz-Landau, 2022).
5. Castiglione, A., Brick, C., Holden, S., Miles-Urdan, E. & Aron, A. R. Discovering the psychological building blocks underlying climate action: A longitudinal study of real-world activism. *R. Soc. Open Sci.* 9, 210006 (2022). https://doi.org/10.1098/rsos.210 006
6. Hamann, K. R. S., von Agris, A.-S. & Markus, L. Investigating the predictors of collective action intensity and health. https://osf.io/preprints/psyar xiv/qev28v1 (2023).
7. van Zomeren, M., Postmes, T. & Spears, R. Toward an integrative social identity model of collective action: A quantitative research synthesis of three sociopsychological perspectives. *Psychol. Bull.* 134, 504–535 (2008). https://doi.org/10.1037/0033-2909.134.4.504

INDEX

Note: Page numbers in **bold** refer to Tables.